Grazing Research: Design, Methodology, and Analysis

Grazing Research: Design, Methodology, and Analysis

Proceedings of a symposium sponsored by Divisions C-6 and C-3 of the Crop Science Society of America in Anaheim, CA, 28 Nov. 1988. Cosponsors include the American Forage and Grassland Council, Society for Range Management, and American Society of Animal Science.

Editor
G. C. Marten

Editorial Committee
G. C. Marten, *chair*
D. P. Hutcheson
M. E. Riewe

Organizing Committee
M. E. Riewe, *chair*
D. I. Bransby
R. H. Hart
A. G. Matches
F. M. Rouquette, Jr.

Editor-In-Chief CSSA
C. W. Stuber

Editor-In-Chief ASA
G. H. Heichel

Managing Editor
Susan Ernst

CSSA Special Publication Number 16

**Crop Science Society of America, Inc.
American Society of Agronomy, Inc.
Madison, Wisconsin, USA**

1989

Cover Design: Patricia Jeffson

Copyright © 1989 by the Crop Science Society of America, Inc.
American Society of Agronomy, Inc.

ALL RIGHTS RESERVED UNDER THE U.S. COPYRIGHT LAW
OF 1978 (P.L. 94-553)

Any and all uses beyond the limitations of the "fair use" provision of the law require written permission from the publisher(s) and/or the author(s); not applicable to contributions prepared by officers or employees of the U.S. Government as part of their official duties.

Crop Science Society of America, Inc.
American Society of Agronomy, Inc.
677 South Segoe Road, Madison, WI 53711, USA

Library of Congress Cataloging-in-Publication Data

Grazing research: design, methodology, and analysis: proceedings of a symposium/sponsored by Divisions of C-6 and C-3 of the Crop Science Society of America in Anaheim, CA, 28 Nov. 1988: cosponsors include the American Forage and Grassland Council, Society for Range Management, and American Society of Animal Science: editor, G.C. Marten.
    p. cm. -- (CSSA special publication: no. 16)
  ISBN 0-89118-527-5
    1. Grazing--Research--Congresses. I. Marten, G.C. II. Crop Science Society of America. Division C-6. III. Crop Science Society of America. Division C-3. IV. American Society of Agronomy. V. Series.
SF84.86.G73 1989
636.08'4--dc20   89-15864
                                                                             CIP

Printed in the United States of America

CONTENTS

Foreword vii
Preface ix
Contributors xi

1 Objectives of Grazing Trials
    Richard H. Hart and Carl S. Hoveland 1

2 The Relationship of Herbage Mass and Characteristics to Animal Responses in Grazing Experiments
    J.C. Burns, H. Lippke, and D.S. Fisher 7

3 Measurement of Animal Response in Grazing Research
    John A. Stuedemann and Arthur G. Matches 21

4 Measurements of the Plant-Animal Interface in Grazing Research
    S.W. Coleman, T.D.A. Forbes, and J.W. Stuth 37

5 Compromises in the Design and Conduct of Grazing Experiments
    David I. Bransby 53

6 Compromises and Statistical Designs for Grazing Experiments
    J. Wanzer Drane 69

7 Experimental Design and Statistical Inference: Generalized Least Squares and Repeated Measures over Time
    F.G. Giesbrecht 85

8 Time Series, Dynamic Models, and Adaptive Sequential Decisions in Grazing Research
    Donald A. Jameson 97

9 Economic Considerations in Grazing Research
    Lucas D. Parsch and L. Allen Torell 109

10 Issues on Modeling Grazing Systems
    Otto J. Loewer 127

FOREWORD

Grazing of pasture and range lands by livestock is an integral component of livestock production systems in the USA and many other countries. Grazing management includes critical decisions concerning the animals and the plants upon which they feed. Profound long-term effects of grazing can be manifested in the classical fence-row photographs of overgrazed and properly grazed range or pasture. Thus, grazing management strategies influence conservation of soil, water, and biological organisms. There is a call for greater diversity as a component of long-term sustainability of agriculture. A livestock component in sustainable agricultural systems contributes to greater options in crop rotation, utilization of forages, and improvement of soil structure and fertility.

Proper methodology is a critical issue in conducting research to develop production management practices for livestock grazing. The interacting effects of animals and plants of various species, age, and condition pose one of the most complex research design problems in agriculture. It is timely that this topic be given special attention as new problems require solutions and new technology becomes available to the researcher. We compliment the organizers of the symposium that resulted in this book. A modern assessment of the technical challenges in conducting grazing research is presented. We expect this book will stimulate additional development in research design and better understanding on how to conduct, interpret, and report results from this important research area.

CALVIN O. QUALSET, *president*
Crop Science Society of America

EDWARD C. A. RUNGE, *president*
American Society of Agronomy

GORDON C. MARTEN, *editor*

PREFACE

Grazing experiments with livestock are required to define input-output relationships that cannot be provided by laboratory, greenhouse, or small field plot studies. Inputs are considered treatments in the design of grazing experiments and may include grazing systems, stocking rates, kinds of pasture plants or animals, pasture fertilization levels, and other alternatives or combinations. Outputs include animal, plant, and/or economic responses. Almost without exception, grazing experiments constitute mission-oriented research and usually should produce results that are directly or indirectly relevant to the livestock producer.

Conducting a proper grazing study poses a considerable challenge. Grazing experiments are usually expensive in terms of land, livestock, equipment, and time. Thus, questions concerning cost and efficiency are relevant and must be addressed by researcher and research administrator alike.

The major concerns in grazing research are to design an experiment that (i) provides for valid comparisons among treatments, (ii) provides a valid error term for evaluating treatment effects, (iii) explores relevant interactions, (iv) provides the data that are useful in extending the applicability of results, (v) provides a reasonable basis for economic assessment of the worth of selected inputs, and/or (vi) defines the relevant input-output relationships in a manner useful to the producer.

Input-output relationships at the plant-animal interface cannot usually be satisfactorily defined without appropriate measures to characterize the dynamics of the sward as well as the response of animals grazing the sward. These relationships must be understood to facilitate mathematical modelling of the grazing ecosystem.

This special publication encompasses the papers from a symposium held at the annual meeting of the Crop Science Society of America in Anaheim, CA, in November 1988. The symposium and this special publication were cosponsored by the Crop Science Society of America, the American Society of Agronomy, the American Forage and Grassland Council, the Society for Range Management, and the American Society of Animal Science.

The participants in the symposium were all experts in their subject. Thus, this special publication will be useful to those involved in the design and execution of grazing research in pasture and range and the interpretation of data derived from grazing experiments. It should also be helpful to those who have responsibility for the review, publication, and/or application of the results from such research.

The Organizing Committee

M. E. RIEWE, *chair*
Texas A&M Agricultural Research Station
Angleton, Texas

D. I. BRANSBY
Auburn University
Auburn, Alabama

R. H. HART
USDA-ARS
Cheyenne, Wyoming

A. G. MATCHES
Texas Tech University
Lubbock, Texas

F. M. ROUQUETTE, JR.
Texas A&M Agricultural Research and Extension Center
Overton, Texas

CONTRIBUTORS

David I. Bransby	Associate Professor, Department of Agronomy and Soils, Auburn University, Auburn University, AL 36849
J. C. Burns	Plant Physiologist, USDA-ARS and Departments of Crop Science and Animal Science, North Carolina State University, Raleigh, NC 27695
S. W. Coleman	Research Animal Nutritionist, USDA-ARS, Forage and Livestock Research Laboratory, El Reno, OK 73036
J. Wanzer Drane	Professor of Biostatistics, Department of Epidemiology and Biostatistics, School of Public Health, University of South Carolina, Columbia, SC 29208
D. S. Fisher	Plant Physiologist, USDA-ARS and Department of Crop Science, North Carolina State University, Raleigh, NC 27695
T. D. A. Forbes	Grazing Ecologist, Texas A&M University, Uvalde, TX 78801
F. G. Giesbrecht	Professor of Statistics, Department of Statistics, North Carolina State University, Raleigh, NC 27695
Richard H. Hart	Range Scientist, USDA-ARS, High Plains Grasslands Research Station, Cheyenne, WY 82009
Carl S. Hoveland	Terrell Distinguished Professor, Agronomy Department, University of Georgia, Athens, GA 30602
Donald A. Jameson	Professor, Department of Range Science, Colorado State University, Fort Collins, CO 80523
H. Lippke	Associate Professor, Texas Agricultural Experiment Station, Angleton, TX 77516
Otto J. Loewer	Professor and Head, Biological and Agricultural Engineering Department, University of Arkansas, Fayetteville, AR 72701
Arthur G. Matches	Thornton Professor, Department of Agronomy, Horticulture, and Entomology, Texas Tech University, Lubbock, TX 79409
Lucas D. Parsch	Associate Professor, Agricultural Economics, University of Arkansas, Fayetteville, AR 72701
John A. Stuedemann	Animal Scientist, USDA-ARS, Southern Piedmont Conservation Research Center, Watkinsville, GA 30677
J. W. Stuth	Professor, Department of Range Science, Texas A&M University, College Station, TX 77843
L. Allen Torell	Associate Professor, Department of Agricultural Economics, New Mexico State University, Las Cruces, NM 88003

1 Objectives of Grazing Trials

Richard H. Hart
USDA-ARS
High Plains Grasslands Research Station
Cheyenne, Wyoming

Carl S. Hoveland
University of Georgia
Athens, Georgia

ABSTRACT

Objectives of grazing trials parallel the objectives of grazing, which are improvement or maintenance of forage production, efficient use of forage produced, and sustained high forage and animal production. In turn, trial objectives and biological and financial constraints determine the variables to be measured and how they are to be measured, the design and analysis of the trial, and interpretation of the results.

Grazing trials are established to provide guidelines for practical grazing management and/or basic information on biology of pasture and range ecosystems. The objectives of grazing trials parallel the objectives of grazing, which Heady (1975) listed as follows:

1. Improvement or maintenance of forage production and range or pasture condition
2. Efficient use of forage produced
3. Sustained high carrying capacity and animal production

The objectives of a grazing experiment in turn determine the variables to be measured and how they should be measured. Morley (1978) pointed out that measurement of animal performance (gain, reproduction, product quality, etc.) is usually the central objective of a grazing experiment, but measurements of animal production alone are seldom adequate to provide an understanding of the complexities of a grazing situation. Thus, information may be sought on response of plants and plant communities to grazing, animal behavior, soil and hydrologic responses, economics, or any combination of these and other variables.

Copyright © 1989 Crop Science Society of America and American Society of Agronomy, 677 S. Segoe Rd., Madison, WI 53711, USA. *Grazing Research: Design, Methodology, and Analysis*, CSSA Special Publication no. 16.

If animal performance is a major objective but there are pasture variables that interact (for example, species composition, fertility, or grazing systems), then measurements of forage quantity and quality, soil water and fertility parameters, and weather may be desirable. If plant responses to grazing are major objectives and the animal is used as a tool to apply grazing pressure, more detailed measurements of plants are needed. Measurements from this kind of experiment also provide guidelines for realistic simulation of grazing in clipping and greenhouse experiments (Hart & Balla, 1982; Stroud et al., 1985). On the other hand, if animal response to supplemental feed or pest control is the objective and pasture is a uniform component of all treatments, intensive sampling of pastures for herbage yield or quality may be a waste of time and money.

ANIMAL PERFORMANCE

Animal performance under grazing depends on the intrinsic productive capacity of the animals, forage production and quality, treatment, and stocking rate (SR, animals per unit area) or grazing pressure (GP, animals per unit of forage produced). Production at the economic optimum or most profitable SR is the most useful measure of treatment effect on animal production. Because of the variability among years in forage production, livestock producers can fix SR but not GP at the beginning of the grazing season. However, GP at a particular time can be adjusted as the season progresses. Stability of returns and pasture or range condition also must be considered.

Determination of the economic optimum SR requires calculation of functions showing the relationship of animal performance to increasing SR. Gain or conception rate usually fit a model with a constant level of performance up to a critical level, then a linear decline as GP increases beyond the critical level. Therefore, a grazing experiment to measure animal response should include a minimum of two GP values on the declining portion of the response curve; a better design includes four GP values, one below the critical GP and three above it. Once the response curve is calculated, simple equations can be used to calculate the most profitable SR or GP (Hart, 1978; Hart et al., 1988a, b). It should be emphasized that SR or GP is a treatment, not a response. Such statements as "Carrying capacity was increased 30% by Treatment A" are meaningless without reference to a level of animal gain, milk or wool production, or reproduction.

PLANT RESPONSES

The grazing experiment on perennial pasture or range should indicate whether the SR that produces the greatest short-term return will provide long-term stability of forage production and the grazed plant community. Excessive numbers of animals, in combination with climatically induced variations in forage growth rate, can affect sward responses such as tillering so that production levels are impaired for a considerable period of time (Grant et al., 1985; Hart & Norton, 1988). The grazing animal exerts four major effects on a pasture.

1. Defoliation of herbage reduces photosynthetic capacity and may reduce root development, carbohydrate storage, and N_2 fixation
2. Selectivity of plant parts and plant species in a mixed stand will affect relative productivity and persistence of the species present and invasion of undesirable species
3. Trampling damages plant tissue, increases soil bulk density, and slows water infiltration
4. Excretion concentrates urine and dung in small areas and affects plant palatability and nutrient cycling

Measurement of these responses may explain why observed differences in production per animal or per unit area occur, and determine how far the results of an experiment can be extended to other areas (Taylor, 1985).

MEASUREMENT TECHNIQUES

Techniques of measurement are covered in detail in other chapters of this publication, but will be presented here in outline because objectives may determine techniques. When measuring animal gain, weighing more often than every 4 wk may not improve precision. Weighing after shrinking may be as efficient in reducing variance as weighing for 3 successive days without shrinking. Automated identification, weighing, and recording of weights shows promise for reducing labor and variance, but at considerable cost for purchase and maintenance of complex equipment.

Intensive grazing systems involving rotation of animals through several pastures may be beneficial without increasing forage production or conversion efficiency if they promote more uniform distribution of grazing. Distribution and activity of grazing animals may be observed directly or measured with mechanical or electronic devices. Direct observation is most reliable but labor-intensive; instruments to make these measurements are expensive and prone to failure at critical moments, but these problems should diminish with further development. Output of measuring devices requires careful interpretation; for example, pedometers do not measure distance traveled, but count steps, which are multiplied by an assumed constant distance per step. Thus, if the objective is to estimate trampling impact, pedometers are suitable; if the objective is to estimate distance traveled, observation and mapping are better.

The plant-animal interface is an area often neglected in grazing research, but it may reveal the reasons why specific grazing management techniques are or are not successful. Plant parts, whole plants, plant communities, and sward structures can be measured or photographed at intervals to determine frequency and intensity of grazing. This information can be used to design simulated grazing experiments or can be incorporated as components of grazing simulation models.

Forage intake is a major factor determining animal response and is most difficult to estimate. Most estimates are based on digestibility of the diet and excretion of indigestible diet fractions and markers, although more direct measures of intake have been attempted. Diet digestibility and selectivity of plants and plant parts should be estimated from samples taken with esophageally fistulated animals, because diets selected are seldom the same as forage on offer. Rare earths may be better indicators of fecal volume than is chromic oxide. A nearly constant level of indicators in the digestive tract may be achieved with impregnated paper, infusion pumps, or time-release capsules.

Forage quality may be measured by in vivo or in vitro digestibility, wet chemistry, or near-infrared reflectance spectroscopy. Differences in composition may be difficult to interpret; acid detergent fiber (ADF) and lignin are associated with digestibility and neutral detergent fiber (NDF) with intake, but relationships are variable and are particularly difficult to establish in complex plant communities. Forage antiquality components such as alkaloids, fungal endophytes, or tannins may be much more important than conventional measures of forage quality. For instance, most of the older grazing studies on tall fescue pastures showed poor animal performance, but removal of the fungal endophyte resulted in gains similar to those on other cool-season perennial grasses (Hoveland et al., 1983).

Forage production and availability may be measured by clipping or double-sampling with a combination of clipping and capacitance meter, beta attenuation meter, weighted disc, or visual estimation. Stationary cages may be used to estimate peak biomass; movable cages may be used to estimate seasonal dynamics of production. Frequency, cover, and density of individual species allow identification of trends in botanical composition. Root dynamics and age structure of plant populations are difficult to measure but may provide early indicators of population trends.

Soil bulk density and water infiltration measurements evaluate trampling effects. An infiltrometer cannot measure effects of surface roughness on water retention and infiltration. Rainfall simulators and runoff plots, which measure these effects, are time-consuming and expensive. Measurements of nutrient concentrations, distribution of urine and feces, and microbial activity evaluate nutrient cycling. On rangelands, it may be important to consider impacts on wildlife food and cover and on amenity values.

DESIGN, MANAGEMENT, AND APPLICATION

Few grazing researchers are fully trained in experimental design and data analysis. Alternatives to complete replication for estimating variance, selection of proper error terms, and proper pasture size and animal numbers must be considered. Covariance analysis can increase the information from a study containing three or more stocking rates or grazing pressures without land replication (Riewe, 1961). Instead of acquiring statistical competence, it may be more efficient to work with a statistician who understands the biological and financial as well as the statistical limitations to experimental design.

To achieve the objectives of a grazing trial, each treatment should be managed for maximum performance. It is appropriate to be flexible and not insist on uniform management, provided management differences can be justified biologically. If Treatment A permits earlier turnout than Treatment B, this should be done; however, all management variables should be properly recorded so that results can be verified by others or compared to previous research or to producer experience.

Finally, one should consider the application of the results of grazing research. Design, interpretation, and presentation should be influenced by whether the results are to be used to develop management alternatives, to be incorporated as components of simulation models, or to provide a basis for further research.

REFERENCES

Grant, S.A., J. King, and G.T. Barthram. 1985. The role of sward adaptations in buffering herbage production responses to management manipulation. p. 1114–1116. *In* Proc. 15th Int. Grassl. Congr., Kyoto, Japan. 24–31 August. The Natl. Grassl. Res. Inst., Nishi-narino, Japan.

Hart, R.H. 1978. Stocking rate theory and its application to grazing on rangelands. p. 550–553. *In* D.N. Hyder (ed.) Proc. 1st Int. Rangel. Congr., Denver, CO. 14–18 August. Society for Range Management, Denver, CO.

Hart, R.H., and E.F. Balla. 1982. Forage production and removal from western and crested wheatgrasses under grazing. J. Range Manage. 35:362–366.

Hart, R.H., and B.E. Norton. 1988. Grazing management and vegetation responses. p. 493–525. *In* P.T. Tueller (ed.) Vegetation science applications for rangeland analysis and management. Kluwer Academic Publ., Dordrecht, Netherlands.

Hart, R.H., M.J. Samuel, P.S. Test, and M.A. Smith. 1988a. Cattle, vegetation and economic responses to grazing systems and grazing pressure. J. Range Manage. 41:282–286.

Hart, R.H., J.W. Waggoner, Jr., T.G. Dunn, C.C. Kaltenbach, and L.D. Adams. 1988b. Optimal stocking rate for cow-calf enterprises on native range and complementary improved pastures. J. Range Manage. 41:435–441.

Heady, H.F. 1975. Rangeland management. McGraw-Hill, New York.

Hoveland, C.S., S.P. Schmidt, C.C. King, Jr., J.W. Odom, E.M. Clark, J.A. McGuire, L.A. Smith, H.W. Grimes, and J.L. Holliman. 1983. Steer performance and association of *Acremonium coenophialum* fungal endophyte on tall fescue pasture. Agron. J. 75:821–824.

Morley, F.H.W. 1978. Animal production studies on grassland. p. 103–162. *In* L. 't Mannetje (ed.) Measurement of grassland vegetation and animal production. Commonw. Agric. Bur. Bull. 52. Hurley, Berkshire, England.

Riewe, M.E. 1961. Use of the relationship of stocking rate to gain of cattle in an experimental design for grazing trials. Agron. J. 53:309–313.

Stroud, D.O., R.H. Hart, M.J. Samuel, and J.D. Rodgers. 1985. Western wheatgrass responses to simulated grazing. J. Range Manage. 38:103–108.

Taylor, J.A. 1985. The animal factor in pasture studies. p. 1140–1142. *In* Proc. 15th Int. Grassl. Congr., Kyoto, Japan. 24–31 August. The Natl. Grassl. Res. Inst., Nishi-narino, Japan.

2 The Relationship of Herbage Mass and Characteristics to Animal Responses in Grazing Experiments[1]

J. C. Burns

USDA-ARS
North Carolina State University
Raleigh, North Carolina

H. Lippke

Texas Agricultural Experiment Station
Angleton, Texas

D. S. Fisher

USDA-ARS
North Carolima State University
Raleigh, North Carolina

ABSTRACT

Evaluation of forage species and management factors with grazing animals requires judicious planning and clearly stated objectives. In these macro-experiments, the cost of land, animals, and labor can restrict the objectives of the study. Further, measurements taken during a trial should not be a consequence of resources; rather, they should be selected to meet the objectives. Animal response data are collected, but frequently little or no pasture characterization occurs, so potential explanatory power of the variables that influence animal response is negated. Although measurements of animal response and pasture productivity have producer utility, they can only be discussed in vague terms and, consequently, neither knowledge nor understanding of pasture and animal interrelationships are advanced. This chapter outlines the importance of estimating herbage mass (kg ha^{-1}) in grazing experiments. Herbage mass is one of four pasture measurements recommended in the conduct of all grazing experiments. An estimate of herbage mass provides basic information from

---

[1] Cooperative investigation of the USDA-ARS and the North Carolina Agric. Res. Serv., Raleigh, NC. Paper no. 11927 of the Journal Series of the North Carolina Agric. Res. Serv., Raleigh, NC 27695-7643 and Technical Article no. TA 24351, Texas Agric. Exp. Stn.

Copyright © 1989 Crop Science Society of America and American Society of Agronomy, 677 S. Segoe Rd., Madison, WI 53711, USA. *Grazing Research: Design, Methodology, and Analysis*, CSSA Special Publication no. 16.

which some understanding of the plant animal interaction can be derived. Justification for estimating herbage mass, methods of estimation, its use in reporting animal response data, and its relationships with animal response are discussed. The other three variables recommended for routine measurements are green (nonsenescent) leaf mass (kg ha^{-1}), diet quality, and herbage desnity (kg ha^{-1} cm^{-1}). Additional desirable explanatory measurements are suggested and prioritized.

In modern nutritional research, the feed offered to an animal is carefully measured and controlled to maximize uniformity of consumption. Similar monitoring of the diet is mimicked in commercial feed lots. In those cases, careful monitoring of the feed permits control of the animal's diet and, within the range of animal-to-animal variation, feeding to meet the animal's requirement for a specific animal response. In grazing research, the diet of the animal is influenced only indirectly by humans. Here, the research focus is not on a controlled static diet but rather on the description of a dynaimc diet. The research objectives should include measurements of pasture characteristics that relate to ease of prehension. Intake of the desirable plant parts is the primary determinant of production by the grazing animal. Herbage mass (Hodgson, 1979), synonymous with available forage or available pasture (Blaser et al., 1986), is the measure most often associated with ease of prehension.

Historically, grazing experiments in the USA have been production-oriented and beef steers (*Bos* spp.) have been the primary experimental animal (Burns and Standaert, 1985). These studies have provided producers with gain per animal, expressed frequently as average daily gain (ADG) and animal gain per hectare, which are valuable guidelines for selection of species, fertility levels, etc. However, measurements that relate to ease of prehension have generally been lacking or inadequate. This leaves the experimenter with little insight into why treatments differed. The high cost of grazing experiments is difficult to justify if measurements taken during a 2- to 5-yr study are inadequate to determine why pasture treatments differed. We examine the importance of estimating herbage mass (HM), and its relationship to animal responses, and we recommend additional variables for consideration in the conduct of grazing experiments.

HERBAGE MASS

Need and Expression

Stocking rate (animals ha^{-1}) or some estimate of pasture productivity (i.e., animal days ha^{-1}) is generally reported in grazing studies along with ADG and gain per hectare. Stocking rate is only indirectly related to ADG and gain per hectare, however, because sward canopy (Hodgson, 1979) or HM does not enter into its definition. In studies with a range of stocking rates, the duration of the experiment alters the observed relationship. Consequently, comparable stocking rates within or among experiments may result in widely different ADG and gain per hectare if HM and sward canopy differ sufficiently. Conversely, differing stocking rates within or among experiments may result in similar animal responses.

Because stocking rate, when applied in multiple increments, has such a profound impact on HM, it generates relationships with ADG and gain per hectare that are predictable and highly significant (Jones, 1974; Mott, 1960; Hart, 1978). This is exaggerated if the duration of the experiment is adjusted based on HM for any one stocking rate, whether the decision is a priori or instantaneous. Although these relationships are in demand by producers and economists, they provide little explanatory power unless accompanying measurements allow an understanding of the biology.

Herbage mass (kg ha^{-1}), or an expression that reflects knowledge of HM (such as herbage allowance; Hodgson, 1979), is a measurement essential to the accurate interpretation of animal response in all experiments. A single measurement provides an instantaneous estimate of herbage; multiple measurements permit tracking the changes that occur during a grazing period or for a growing season. Measurements of HM taken often (weekly) are most useful, but even taken infrequently (monthly) they provide essential information about the treatments.

The effort and cost of obtaining HM is warranted because of its multiple roles. First, it is part of the definition of the forage treatment. In grazing experiments using fixed, continuous stocking, HM can vary widely from excess to intermittently deficient. But, in variable (put-and-take), continuous stocking, the HM is generally maintained within some bounds of a predetermined level by adjusting stocking rate (Wheeler et al., 1973). Second, HM permits interpretation of the stocking rate (i.e., herbage allowance can be evaluated). Third, HM provides a guide to pasture conditions throughout the experiment. Measurement of HM can characterize the dynamics of the treatments and provide a better understanding of the animal response obtained. Finally, HM provides a basis for other calculations such as growth, utilization, and deterioration.

Measurement and Use of Data

The most accurate way to measure HM is by harvesting the total plot at the soil surface. In small plot experimentation, much of the plot area can be harvested and weighed. Generally, harvesting is done to a predetermined stubble height. In grazing experiments, plot (paddock) size is increased several orders of magnitude compared with small plots, causing increased sampling costs (time and labor). In addition, disturbance of animal grazing behavior must be avoided and maximum HM must be left for grazing. These both dictate that subsampling be practiced.

Methods for estimating HM can be classified as direct (destructive sampling) or indirect (nondestructive estimates). Several direct and indirect methods have already been described, assessed, and evaluated for temperate pastures (Frame, 1981; 't Mannetje, 1978; Meijs et al., 1982) and range lands (Shepherd, 1962). These methods also have been reviewed for tropical pastures (Shaw et al., 1976). Consequently, only mention of the methods and their potential will be considered.

Direct Sampling

Subsampling directly ranges from use of a quadrat of a specific frame size hand-harvested at the soil surface to harvesting long, narrow strips mechanically at a specific stubble height. The number, type, size, and location of the quadrats will depend on the accuracy of the estimates needed (Carter, 1962; McIntyre, 1978; Shepherd, 1962). Herbage mass can be reported on a dry matter or organic matter basis, depending on degree of soil contamination.

Stubble height must be consistent and below the grazing height. If differential stubble height is required to maintain stands of different species, it may be necessary to restrict sampling of an area to once a season or to hand-sample a small area at the soil surface within the initial cut and add the estimated residual mass to the HM estimated from the higher stubble ('t Mannetje, 1978). A sward's total HM can only be estimated by cutting at the soil surface. Frame (1981) recommended this approach when studying sward characteristics, herbage production, and animal responses to HM. Direct sampling is the desired method of estimating HM, especially when dry matter intake is to be estimated (Meijs et al., 1982). Direct sampling, however, has limitations that can be overcome by indirect methods.

Indirect Estimates

Indirect estimates of HM in grazing experiments can be justified and are more likely to be used in demonstrations or by producers than are direct estimates. Indirect methods of estimating HM have been classified as visual, height and density, and assessment of various nonvegetative attributes ('t Mannetje, 1978). Recommended use of these methods (Frame, 1981; 't Mannetje, 1978) depends on the need to do the following:

1. Reduce the cost of sampling through reduced labor, time, and equipment.
2. Obtain multiple measurements on large areas (grazing paddocks) or in remote sites.
3. Avoid harvesting large proportions of the HM.
4. Minimize the disturbance of the canopy.
5. Negate the stubble problem associated with cutting height.
6. Rank treatments in trials where large comparative differences exist.
7. Acquire only a relative index of HM.
8. Increase precision or accuracy of direct measures where large variation exists and a large number of estimates are needed.

Estimates from each of the three indirect-method classes have been discussed ('t Mannetje, 1978; Frame, 1981; Meijs et al., 1982). Although either height or density can be estimated alone (i.e., height by ruler or density by some score), and are useful, a device like the weighted disc (Powell, 1974) or any of the rising or falling disc meter methods (Bransby et al., 1977; Earle and McGowan, 1979) give an integrated reading of both height and density. The nonvegetative methods include capacitance meters, radioisotope attenuation, spectral analyses, etc., and have been considered by 't Mannetje (1978) and more recently by Frame (1981). The capacitance meter has advanced (Toledo et al., 1980) to a single probe design (Vickery et al., 1980) and is marketed as the pasture probe (Crosbie et al., 1987).

All indirect methods require calibration when quantification (kg ha^{-1}) is desired. Calibration employs a double-sampling technique in which indirect readings are associated with actual HM obtained by direct sampling; the relationship between the two (regression equation) are used as the basis for predicting direct values (Frame, 1981). Indirect estimates have greater error and are more liable to bias than direct methods (Meijs et al., 1982). The error problem with indirect sampling can be offset with increasing numbers of readings per unit area. Bias remains a dilemma, but can be reduced by frequent calibration.

The indirect methods are sufficiently useful and inexpensive that they can be justified in every grazing experiment. Consequently, no trial should be conducted without HM estimates on a reasonable time schedule.

Reporting Data

Presentation of HM data taken during the season will depend on the frequency of sampling. If weekly values are available for each treatment, the plotting of the trend with confidence intervals should be most useful. If, however, HM values are limited, the data might be most useful if averaged for uniform segments or periods of the grazing season. This may correspond to the seasons, i.e., spring, summer, or fall (Burns et al., 1984). Estimates within each period should then be tested by analyses of variance and the reported means should be accompanied by an estimate of variation (SE).

Herbage mass and its characteristics should be integrated into the discussion of the animal responses as shown by Fisher et al. (1987). This provides primary data on the pasture treatments and enables proper interpretation of ADG and gain per hectare. However, animal response and HM may not always be highly associated across treatments because they are indirectly related.

HERBAGE MASS AND ANIMAL RESPONSE

The relationship of HM to animal performance is frequently percieved as cause-and-effect when it should be more appropriately viewed as an association. Herbage mass forms the bounds from which the animal selects its diet. Animal performance is a direct effect of the quality and quantity of the prehended herbage. The diet's effectiveness in driving animal response is modified by the animal's ability to digest and convert the ingested dry matter into useful nutrients.

At some HM, generally above 2 Mg ha^{-1} for temperate swards (Allden and Whittaker, 1970) and 4 Mg ha^{-1} for subtropical forages (Forbes and Coleman, 1987), animals can readily select a diet of their choice in an acceptable time of grazing (6–9 h). At these levels, HM is mainly descriptive and a low and variable correlation can be expected between HM and animal performance. In selecting the diet of their choice, animals may closely graze a small percentage of the paddock (spot grazing) and ignore the rest. Animals return to these spots and as a consequence obtain a diet higher in quality. However, performance may be moderate, regardless of HM, because dry matter intake may be limited from the preferred areas. In essence, animals have by choice shifted the treatment to one of a much lower herbage mass and lower herbage allowance (essentially a higher stocking rate) without regard to the experimenter's designation. Under such conditions, HM describes the *intended* treatments, but is essentially useless in explaining animal performance. Reporting an estimate of HM variation (SD) will indicate spot grazing.

When HM is declining, the association between HM and animal response increases. Although dry matter intake may be temporarily maintained by increased grazing time and number of bites, bite-size actually declines (Arnold, 1981). The increased association between HM and animal response results because intake rate may be reduced fourfold, whereas time spent grazing will rarely double before it reaches an upper limit of 12 to 14 h (Freer, 1981; Allden and Wittaker, 1970; Forbes and Coleman, 1987). In addition, the less-preferred portion of the HM will make up a greater proportion of the diet, causing diet quality to become similar to the mean quality of the HM.

At low HM, diet selection is still operative, but then it is dominated by the hunger drive and all of the grazeable HM is consumed (to some minimal residue). Under continuous exposure to a low HM, ease of prehension may be compromised sufficiently to reduce intake and animal response, even though diet quality may be high. Consequently, when HM levels are varied from low to high, the correlation between animal daily response and HM is high ($r = 0.97$; Marsh, 1979) with intake reflecting HM. This response occurs when ranges of stocking rates are evaluated, giving the expected relationship of decreased ADG with increasing stocking rate. This relationship holds even though HM may be limiting only toward the end of the grazing season (Hart, 1978). Ease of prehension and, therefore, intake are conditioned not only by the extent to which preference and selection occurs within any HM, but also by physical presentations of the sward canopy. This aspect is considered further by Coleman et al. (1989) (see Chapter 4 of this publication).

Herbage mass and animal response relationships shift under declining and limited HM from an indirect association to one of increasing cause-and-effect as the HM represents the diet of the animal more and more closely. Inclusion of HM estimates when reporting data from grazing experiments provides an understanding of the extent to which this shift has occurred.

HERBAGE MASS, QUALITY, AND ANIMAL RESPONSE

Within the association between HM and animal response are the dynamics of HM quality and diet quality. An estimate of HM quality (e.g., in vitro dry matter disappearance) provides only a mean value for the sward canopy, which generally declines with increasing maturity and/or HM. When HM is moderate to abundant, the quality of the diet will be higher than the mean of the sward canopy, because animals selectively graze green vs. dead herbage and green leaf from the upper portion of the canopy vs. stem (Arnold, 1981). The differential is smaller if leaf and stem quality are similar or if leaf prehension is difficult. Selective grazing frequently causes HM quality and diet quality to be poorly associated, resulting in low correlations between HM and animal response (Burns et al., 1984). In these cases, HM quality is only descriptive and provides little information about the resulting quality of the diet and subsequent animal response.

In situations of limited HM, its mean quality should nearly equal diet quality because of diminished selective grazing, reduced maturity of the sward canopy, and a higher proportion of the HM being composed of new growth as well as consumption (allowing some minimal stubble) of the total HM. However, subsequent animal response may not reflect the potentially high quality of the diet if HM is sufficiently limited to reduce intake. In this case, HM quality and diet quality are again descriptive and not highly associated with animal response. Controlling HM to assure that intake is not limited and that most of the newly grown HM is being consumed, however, may cause quality of HM and diet to be correlated and to show a relationship with animal response. This situation is likely to occur when temperate species such as Kentucky bluegrass (*Poa pratensis* L.), ryegrass (*Lolium perenne* L.), or tall fescue (*Festuca arundinacea* Schreb.) are continuously grazed to maintain 5 to 10 cm. Here HM is predominantly leaf. This same relationship can hold with the subtropical grass, switchgrass (*Panicum virgatum* L.), if continuously defoliated at 10 to 15 cm. Regrowth is primarily leaf tissue and relished by the animal (Burns et al., 1984). Harvesting HM to an 8- to 10-cm stubble causes the grazing and HM zones to be very similar and, consequently, the quality of both the HM and diet become similar. When comparing forages of similar productivity that differ in nutritive value (including antiquality factors), experimenters utilizing the variable (put-and-take) method of grazing can control HM and should expect higher ADG but lower animal days per hectare on the better pasture. Experimenters utilizing a range of fixed stocking rates should expect higher ADG on the better pastures at low stocking rate(s), but lower ADG on the better pasture at the higher stocking rate(s). The latter results from increased intake, decreased leaf area, and probably reduced growth. The ADG is likely to be less responsive to stocking rate when pastures are of low nutritive value. In all cases, HM is needed for interpretation. Presentation of HM estimates thus provides insight into the conditions of the experiment, the likely composition of the diet, and the reason for the animal response obtained.

HERBAGE MASS AND ANTIQUALITY FACTORS

Fiber (structural) constituents that negatively influence animal response and nonstructural constituents that alter grazing behavior or cause deleterious effects on animal response have been classified as antiquality factors. These factors have been discussed for both grasses and legumes (Matches, 1973; Burns, 1978; Marten, 1981; Hegarty, 1982; Burns, 1985; Minson and Hegarty, 1985) and are not the focus here. Of interest is the value of HM for indicating when secondary compounds are operative.

Easily detected are factors that cause acute, as opposed to chronic, responses (Burns, 1985). In acute cases, animals exhibit obvious symptoms or death. In chronic cases, deleterious compounds may alter grazing behavior and subsequently diet composition and daily intake, or they may disrupt normal animal physiology or occur in only sufficient concentrations to cause submaximal animal performance. Generally their action is subtle, requiring extended exposure or specific environmental or animal physiological conditons, or both, before animal symptoms are observed, if at all. Frequently their occurrence is not considered.

Both reed canarygrass (*Phalaris arundinacea* L.) and tall fescue are recent examples. In the former case, laboratory estimates of quality were similar between clones with high and low palatability scores, yet animal performance was positively associated with palatability. Grazing trial results implicated a differential intake between high and low alkaloid pastures. Consequently, relative HM changes during grazing and associated animal responses were valuable in substantiating negative influences of the alkaloids on animal daily dry matter intake and an eventual association with physiological disturbance in the animal (Marten et al., 1976). Herbage mass estimates were not reported in a similar situation where the presence of a fungal endophyte in tall fescue was attributed to poor animal daily gain (Hoveland et al., 1980). Because the presence of the endophyte and reduced dry matter intake are associated, similar HM and ADG relationships would be expected as noted for pastures of different nutritive value when grazed by either the variable or multiple, fixed stocking methods discussed in the latter portion of the previous section. Reporting the HM would have been valuable in arriving at and substantiating conclusions.

In much of the present grazing research, HM estimates are inadequate, either not being taken or being taken but not reported. These estimates result in an inadequacy in estimating the magnitude of animal response expected from the treatments being compared. Herbage mass estimates and calculated changes in HM, along with observations of animal health or condition and expected performance, would aid in detecting the occurrence of deleterious herbage compounds as opposed to simply recording animal weight responses. These examples support the need for HM estimates in grazing experimentation, especially in the evaluation of introduced or improved cultivars (Jones and Hegarty, 1981) of genera known to contain deleterious compounds.

PASTURE AND ANIMAL MEASUREMENTS NEEDED IN GRAZING EXPERIMENTS

Grazing experiments generally include objectives that involve between or among treatment comparisons of gain per animal and per hectare. Although ADG is frequently ascribed to the quality of the pasture and gain per ha to treatment (pasture) productivity, these are complex responses that cannot be separated and thought of simply as quality vs. productivity—they are influenced by a number of pasture and animal factors. For example, ADG may be poorly associated with pasture quality, whereas weight gain per hectare may not estimate total productivity. The latter requires estimation of maintenance and gain requirements for a particular animal type, condition, weight, and rate of weight change (Peterson and Lucas, 1968). Even when total digestible nutrients or energy harvested per hectare are determined, productivity may be biased by the degree of forage utilization.

The need exists for pasture measurements that characterize differences among treatments in ease of prehension. Such measurements are categorized as being either pasture, animal, or pasture-animal types:

Measurements for Characterization of Pasture

1. *Herbage mass* (kg ha^{-1}). Direct measurements (yield samples) are most useful but indirect estimates are valuable and a minimum requirement for grazing trials. Samples should be drawn for botanical composition if appropriate.
2. *Green leaf mass* (kg ha^{-1}). Senescent or dried material and stems must be removed from the yield sample or adjusted in the estimate.
3. *Pasture growth rate* (kg ha^{-1} d^{-1}). Estimates obtained from herbage enclosures using methods noted in item 1.
4. *Pasture height* (cm). Together with HM (kg ha^{-1}), pasture height provides an estimate of herbage density (kg ha^{-1} cm^{-1}) and an index of sward canopy structure.
5. *Leaf area index* (leaf area m^{-2}).

Measurements for Characterization of Animal Response

6. *Grazing time* (h d^{-1}).
7. *Bite size* (g). This reflects ease of prehension (Stobbs, 1973) but is not always directly related to intake rate. Bite size can also be calculated from items 8 and 9.
8. *Bite number* (h^{-1}). These are prehensile bites counted for short periods of time by visual and audible cues (less accurately for long periods by automated event counters). Daily totals can be calculated by knowledge of item 6 (grazing time). Bite number should not be confused with bite numbers reported in the early literature, which were actually number of mouth movements (chews plus bites plus other).

9. *Dry matter intake* (kg d^{-1}). This will generally be determined by indirect marker techniques in which fecal output is estimated. Knowledge of diet quality (item 12) permits intake calculation.
10. *Rumen fill* (% body weight). This calculation is possible when pulse dose markers are used to estimate fecal output.
11. *Digesta retention time* (h). This calculation is possible when pulse dose markers are used to estimate fecal output.

Measurements for Characterization of Pasture/Animal Dynamics

12. *Diet quality*. Samples representing the animal's diet can be obtained either by esophageal collection or by hand plucking and in some cases by cutting. The best laboratory measure of overall quality is in vitro dry matter disappearance.
13. *Intake rate* (g h^{-1}). This is calculated from grazing time and dry matter intake (items 6 and 9) or from bite size and bite number (items 7 and 8).

The degree to which these 13 variables can or should be incorporated into an experiment depends on the objective(s) of the research and the interest one has in understanding the response obtained. Because funds for labor, supplies, equipment, animals, etc., may determine feasibility of measurements taken, recommended measurements that should be routine in all grazing experiments are listed in Table 2-1, Classification A, along with a relative rank in priority and cost. Herbage mass is ranked first because it provides not only a description of the treatment evaluated, but a biological basis for comparison (nonstatistical) of animal responses among grazing experiments. The use of esophageal collection or hand-plucking or cutting of herbage for diet quality estimates may not be interchangeable and requires local evaluation. If measurements in Classification A cannot be obtained, the value of the experimental results is lessened substantially and the experiment may not provide sufficient information to justify the cost of obtaining the animal measurements.

Additional desirable measurements are suggested in Table 2-1 (Classifications B and C) with their priority ranked for the most efficient expenditure of resources to obtain data necessary to understand the dynamics of the resulting animal responses. The routinely recommended measurements in Table 2-1 (Classification A) are pasture-dominant, whereas other desirable ones (Classification B and C) are dominantly animal in nature.

The adoption of these measurements, and their inclusion with animal response data from grazing experiments, will greatly enhance the biological interpretation and consequently, improve the value of grazing research. Further, it would remove the need to use vague qualitative terms such as high or low grazing or stocking intensity or level. We recommend the use of quantitative measurements (such as HM) to characterize treatments when discussing or presenting experimental results, thereby adding clarity to the conduct and design of grazing experiments.

Table 2-1. Classification, priority rank, description, and cost of obtaining pasture, animal, and pasture/animal measurements relative to their explanatory value in grazing experiments.

Classification and priority rank	Measurements Number†	Measurements Type	Description/option‡	Relative cost§
A. Recommended for all grazing experiments:				
1. Herbage mass	1	Pasture	a. Direct (yield samples)	H
			b. Indirect (relative)	L
			c. Indirect, calib. (estimated)	L
			d. (a + b) or (b + c)	H
2. Green leaf mass	2	Pasture	a. Direct (harvest)	H
			b. Visual (score)	L
3. Diet quality	12	Pasture/animal	a. Esophageal	H
			b. Hand pluck	L
			c. Cut	L
4. Herbage density	4	Pasture	a. Meter	L
			b. With option 1a and 1c need height	L
			c. With option 1b need 1a	H
5. Botanical composition	1	Pasture	a. Hand separate	H
			b. Visual score	L
			c. Chemical or physical est.	L
B. Desirable for additional explanatory purposes:				
1. Grazing time	6	Animal	a. Clock (vibracorder)	L
			b. Observed	H
2. Intake	9	Animal/pasture	a. Direct (yield samples)	H
			b. Marker, pulse dose	H
			c. Marker, continuous release	L
3. Intake rate	13	Animal	(Calc. from B.1 and B.2 or C.2 and C.3)	N
4. Pasture growth rate	3	Pasture	a. Direct (enclosures with 1a)	H
			b. Indirect (enclosures with 1b or c)	L
			c. (Calc. from A.1)	N
C. Desirable for maximum explanatory purposes:				
1. Leaf area index	5	Pasture	a. Hand measurement	H
			b. Instrumentation	H
2. Bite size	7	Animal	a. Cannulated animals	H
3. Bite number	8	Animal	a. Visual and audible cues	H
4. Rumen fill	10	Animal	(Calc. from B.2.b)	N
5. Digesta retention time	11	Animal	(Calc. from B.2.b)	N

† Number corresponds to measurements delineated in text.
‡ Calib. = calibrated; calc. = calculated (numbers refer to classification column).
§ H = high, L = low, and N = none.

REFERENCES

Allden, W.G., and I.A.M. Whittaker. 1970. The determinants of herbage intake by grazing sheep: The interrelationship of factors influencing herbage intake and availability. Aust. J. Agric. Res. 21:755-766.

Arnold, G.W. 1981. Grazing behavior. p. 79-102. *In* F.A.W. Morley (ed.) World animal science. B. Grazing animals. Elsevier Science Publ. Co., New York.

Blaser, R.E., R.C. Hammes, Jr., J.P. Fontenot, H.T. Bryant, C.E. Polan, D.D. Wolf, F.S. McClaugherty, R.G. Kline, and J.S. Moore. 1986. Forage-animal management systems. Virginia Agric. Exp. Stn. Bull. 86-87.

Bransby, D.I., A.G. Matches, and G.F. Krause. 1977. Disk meter for rapid estimation of herbage yield in grazing trials. Agron. J. 69:393-396.

Burns, J.C. 1978. Forage quality and animal performance: Antiquality factors as related to forage quality. J. Dairy Sci. 61:1809-1820.

Burns, J.C. 1985. Antiquality factors in temperate legumes in the United States. p. 260-267. *In* R.F Barnes et al. (ed.) Forage legumes for energy-efficient animal production. Proc. Trilateral Workshop, Palmerston North, N.Z. 30 Apr.-4 May 1984. USDA-ARS, Washington, DC.

Burns, J.C., R.D. Mochrie, and D.H. Timothy. 1984. Steer performance from two Pennisetum species, switchgrass and a fescue-'Coastal' bermudagrass system. Agron. J. 76:795-800.

Burns, J.C., and J.E. Standaert. 1985. Productivity and economics of legume-based vs. nitrogen-fertilized grass-based pastures in the United States. p. 56-71. *In* R.F Barnes et al. (ed.) Forage legumes for energy-efficient animal production. Proc. Trilateral Workshop, Palmerston North, N.Z. 30 Apr.-4 May 1984. USDA-ARS, Washington DC.

Carter, J.F. 1962. Herbage sampling for yield: Tame pastures p. 90-101. *In* G.O. Mott et al. (ed.) Pasture and range research techniques. Comstock Publ. Assoc., Ithaca, NY.

Coleman, S.W., T.D.A. Forbes, and J.W. Stuth. 1989. Measurements of the plant-animal interface in grazing research. p. 37-51. *In* G.C. Marten (ed.) Grazing research: Design, methodology and an analysis. CSSA Spec. Publ. 16. CSSA, ASA, Madison, WI (Chapter 4 of this publication).

Crosbie, S.F., B.M. Smallfield, H. Hawker, M.J.S. Floate, J.M. Keoghan, P.D. Enright, and R.J. Abernethy. 1987. Exploiting the pasture capacitance probe in agricultural research. J. Agric. sci. 108:155-163.

Earle, D.F., and A.A. McGowan. 1979. Evaluation and calibration of an automated rising plate meter for estimating dry matter yield of pasture. Aust. J. Exp. Agric. Anim. Husb. 19:337-343.

Fisher, D.S., K.R. Pond, and J.C. Burns. 1987. An interdisciplinary approach to pasture-animal interface research. p. 3-20. *In* F.P. Horn et al. (ed.) Grazing-lands research at the plant-animal interface. Winrock Int. Inst. for Agric. Dev., Morrilton, AR.

Forbes, T.D.A., and S.W. Coleman. 1987. Herbage intake and ingestive behavior of grazing cattle as influenced by variation in sward characteristics. p. 141-152. *In* F.P. Horn et al. (ed.) Grazing-lands research at the plant-animal interface. Winrock Int. Inst. for Agric. Dev., Morrilton, AR.

Frame, J. 1981. Herbage mass. p. 39-69. *In* J. Hodgsen et al. (ed.) Sward measurement handbook. British Grassl Soc., Grassl. Res. Inst., Hurley, Maidenhead, Berkshire, U.K.

Freer, M. 1981. The control of food intake by grazing animals. p. 105-124. *In* F.A.W. Morley (ed.) World animal science. B. Grazing animals. Elsevier Science Publ. Co., New York.

Hart, R.H. 1978. Stocking rate theory and its application to grazing and rangelands. p. 547-550. *In* D.N. Hyder (ed.) Proc. 1st Int. Rangeland Congr. Denver, CO. 14-18 August. Society for Range Management, Denver, CO.

Hegarty, M.P. 1982. Deleterious factors in forages affecting animal production. p. 133-150. *In* J.B. Hacker (ed.) Nutritional limits to animal production from pastures. Proc. Int. Symp., St. Lucia, Queensland, Aust. 24-28 Aug. 1981. Commonw. Agric. Bureaux, Farnham Royal, Slough, England.

Hodgson, J. 1979. Nomenclature and definitions in grazing studies. Grass Forage Sci. 34:11-18.

Hoveland, C.S., R.L. Haaland, C.C. King, Jr., W.B. Anthony, E.M. Clark, J.A. McGuire, L.A. Smith, H.W. Brimes, and J.L. Holliman. 1980. Association of *Epichloe typhina* fungus and steer performance on tall fescue pasture. Agron. J. 72:1064-1065.

Jones, R.J. 1974. The relationship between animal gain and stocking rate. J. Agric. Sci. 83:335-342.

Jones, R.J., and M.P. Hegarty. 1981. Screening pasture plants for possible toxic effects on livestock. p. 237-247. *In* J.L. Wheeler and R.D. Mochrie (ed.) Forage evaluation: Concepts and techniques. Proc. Bilateral Workshop. Armidale, NSW, Aust. 27-31 Oct. 1980. Am. Forage and Grassl. Council, Lexington, KY.

't Mannetje, L. (ed.). 1978. Measuring quantity of grassland vegetation. p. 63-95. *In* Measurement of grassland vegetation and animal production. Bull. 52. Commonw. Agric. Bureaux, Farnham Royal, Bucks, England.

Marsh, R. 1979. Effect of herbage DM allowance on the immediate and longer term performance of young Friesian steers at pasture. N.Z. J. Agric. Res. 22:209-219.

Marten, G.C., 1981. Effects of deleterious compounds on animal preference for forage and on animal performance. p. 225-260. *In* J.L. Wheeler and R.D. Mochrie (ed.) Forage evaluation: Concepts and techniques. Proc. Bilateral Workshop. Armidale, NSW, Australia. 27-31 Oct. 1980. Am. Forage and Grassl. Council, Lexington, KY.

Marten, G.C., R.M. Jordan, and A.W. Hovin. 1976. Biological significance of reed canarygrass alkaloids and associated palatability variation to grazing sheep and cattle. Agron. J. 68:909-914.

Matches, A.G. (ed.). 1973. Anti-quality components of forages. CSSA Spec. Publ. 4. CSSA, Madison, WI.

McIntyre, G.A. 1978. Statistical aspects of vegetation sampling. p. 8-21. *In* L. 't Mannetje (ed.) Measurement of grassland vegetation and animal production. Bull. 52. Commonw. Agric. Bureaux, Farnham Royal, Bucks, England.

Meijs, J.A.C., R.J.K. Walters, and A. Keen. 1982. Sward methods. p. 11-36. *In* J.D. Leaver (ed.) Herbage intake handbook. The British Grassl. Soc., Grassl. Res. Inst., Hurley, Maidenhead, Berkshire, U.K.

Minson, D.J., and M.P. Hegarty. 1985. Toxic factors in tropical legumes. p. 246-250. *In* R.F Barnes et al. (ed.) Forage legumes for energy-efficient animal production. Proc. Trilateral Workshop, Palmerston North, N.Z. 30 Apr.-4 May 1984. USDA-ARS, Washington, DC.

Mott, G.O. 1960. Grazing pressure and the measurement of pasture production. p. 606-611. *In* Proc. 8th Int. Grassl. Congr., Reading, England. 11-21 July. Alden Press, Oxford, England.

Petersen, R.G., and H.L. Lucas, Jr. 1968. Computing methods for the evaluation of pastures by means of animal response. Agron. J. 60:682-687.

Powell, T.L. 1974. Evaluation of weighted disc meter for pasture yield estimation on intensively stocked dairy pasture. N.Z. J Exp. Agric. 2:237-241.

Shaw, N.H., L. 't Mannetje, R.M. Jones, and R.J. Jones. 1976. Pasture measurements. p. 235-250. *In* N.H. Shaw and W.W. Byron (ed.) Tropical pasture research. Bull. 51. Commonw. Agric. Bureaux, Farnham Royal, Bucks, England.

Shepherd, W.O. 1962. Herbage sampling for yield: Natural pasture and range. p. 102-105. *In* G.O. Mott et al. (ed.) Pasture and range research techniques. Comstock Publ. Assoc., Ithaca, NY.

Stobbs, T.H. 1973. The effect of plant structure on the intake of tropical pastures. I. Variation in bite size of grazing cattle. Aust. J. Agric. Res. 24:809-819.

Toledo, J.M., J.C. Burns, and A. Angelone. 1980. Herbage measurements in situ by electronics. 3. Calibration, characterization and field application of the earth-plate forage capacitance meter: A prototype. Grass Forage Sci. 35:189-196.

Vickery, P.J., I.L. Bennett, and G.R. Nicol. 1980. An improved electronic capacitance meter for estimating herbage mass. Grass Forage Sci. 35:247-252.

Wheeler, J.L., J.C. Burns, R.D. Mochrie, and H.D. Gross. 1973. The choice of fixed or variable stocking rates in grazing experiments. Exp. Agric. 9:289-302.

3 Measurement of Animal Response in Grazing Research

John A. Stuedemann
USDA-ARS
Southern Piedmont Conservation Research Center
Watkinsville, Georgia

Arthur G. Matches
Texas Tech. University
Lubbock, Texas

ABSTRACT

Clearly defined and prioritized objectives ultimately determine the measurements required for each grazing experiment. Forage/livestock research ranges from investigation of univariate processes to entire systems of production. Resources often are limiting in forage-animal research; consequently, the experimental design and responses to be measured should be chosen carefully. Response criteria should be selected only if they are relevant to objectives, can be measured with adequate precision, and permit reasonable testing of hypotheses. If these conditions are not met, resources may be wasted.

Forage/livestock research ranges from investigation of univariate processes to entire systems of production. Experimental resources often are limited in grazing trials; consequently, compromises are made in design and measurements. Measurements of animal response may require substantially more resources than are needed for measurements of vegetation (Morley, 1978). Our objective is to discuss criteria and methods for animal response measurement in grazing experiments.

COPING WITH ANIMAL VARIABILITY

Objectives of the experiment should be clearly defined and prioritized to enable selection of the most appropriate response criteria and methods of measurement. Likewise, the precision and accuracy necessary to meet specific objectives should be understood by the researcher. Generally, pastures are considered the experimental units in grazing trials (Morely, 1978; Brown & Waller, 1986; Cook & Stubbendieck, 1986); however, in special cases animals may serve as the experimental units (e.g., when implants of animals grazing common pastures are evaluated). Animal-to-animal variation is usually the greatest source of variation in grazing trials (Petersen & Lucas, 1960). For example, Mott and Lucas (1953) suggested that pasture variation for production per animal generally has a coefficient of variation (CV) around 5%, whereas the corresponding animal-to-animal variation may have a CV of from 10 to 30%.

Replication, numbers of animals per treatment, and length of grazing period influence the magnitude of experimental errors, and properly applied covariance analysis may be useful in identifying and controlling some sources of animal variation in grazing trials (Petersen & Lucas, 1960; Cook, 1986). In addition, increasing replication of treatments and number of animals per treatment, as well as length of grazing period, will generally improve precision and the power of statistical tests. Estimates of variances and consequent sample sizes needed for various measurements of animal response (Johnson & Laycock, 1963; Matches, 1969), diet quality and composition (Obioha et al., 1972; Holechek & Vavra, 1983), and forage intake (Cordova et al., 1978) have been reported. Optimum animal number is the fewest required to obtain a desired level of precision.

RESPONSE CRITERIA AND THEIR MEASUREMENT

Criteria selected for animal response measurement should be relevant to the hypothesis under study. Experiments should extend for an adequate time to enable response differences to occur. Differences expected among treatments will often dictate which response criteria should be selected. Various response criteria have different inherent characteristics and will yield different information, just as method of measurement may influence the relative value of the data.

Measurements can be categorized into two types, absolute or comparative. *Absolute measurements*, such as scale-weights, include those which enable direct comparison to other research. *Comparative measurements*, such as body condition scores, have value within a given experiment, but may have value to other research if specific standards are utilized.

The process of measuring response criteria or obtaining samples should not interfere with the interpretation of the data. One should be able to obtain measurements or samples without influencing the animal's ability to respond to the treatments imposed.

MEASUREMENT OF LIVE WEIGHT

Live weight is the most common and informative measure of animal performance. Animal weight change has both biological and economic implications because animals are normally marketed on a weight basis. With mature animals (usually breeding animals) weight changes are often coupled with other measures of performance and health, which may include milk production, body condition, tissue composition, and measures of reproductive efficiency. Comprehensive reviews of weight measurement and associated errors and bias include those of Matches, 1970; Hughes, 1976; Corbett, 1978; and Hart, 1987.

Live weight may be biased by gut fill and the time of day at which weights are obtained. Time since last defecation or urination, type and availability of forage, eating or grazing pattern, and weather conditions as they might influence water load in pelage or disturb normal animal behavior may also influence weight measurement.

Manual Weighing

Single-Day versus Multiple-Day Weighing

Early work (Lush & Black, 1927) suggested that successive weighing over 3 d was more accurate than 1-d weighing, but the added accuracy of multiple weighings may not always be justified. Subsequent research has generally indicated that single-day weights are preferable (Baker & Guilbert, 1942; Baker et al., 1947; Patterson, 1947; Bean, 1948; Dembiczak et al., 1957; Harris et al., 1959). Patterson (1947) reported that weighing 11 steers (*Bos* sp.) once reduced variance more than weighing 10 steers on 3 successive days. Multiple-day weights are usually recommended only where accuracy of individual animal weights is desired (Koch et al., 1958; Harris et al., 1959).

Shrunk versus Unshrunk Weighing

Hughes (1976) reported that 12 to 23% of an animal's weight is due to gut contents, which may vary due to type and quality of feed, size and age of animal, and weather conditions. For example, a 410-kg steer may drink 18 to 36 kg of water per day, eat 11 kg of dry matter, but when confined overnight, lose 27 kg of live weight.

Mott (1959) reported that failure to account for differential fill will contribute to experimental errors associated with output per animal. Fluctuations in fill appear to be associated with changes in environmental conditions that influence daily grazing patterns (Baker & Guilbert, 1942; Bean, 1948; Dwyer, 1961; Ehrenreich & Bjugstad, 1966). Also, Whiteman et al. (1954) reported that availability of water may contribute to variation in fill.

Treatment of animals prior to fasting may influence fasting weight losses and subsequent gain or loss. Bailey and Bishop (1975) found that steers wintered on cool-season pasture lost live weight more rapidly in the first 8 h of fasting than did steers fed hay; these differences persisted for at least 48 h of fasting. When both groups were placed on spring pasture, the steers wintered on cool-season pasture exhibited greater gain, which was attributed to lower gut fill before the steers were placed on spring pasture. Bass and Duganzich (1980) reported that cattle lose more live weight during fasting in spring than in summer. Goodchild (1985) found that gut fill was greater during the dry season than during the wet season, and that the proportion of live weight lost during fasting was greater during the wet season. Matches (1981) also reported seasonal differences in the amount of gut fill in that fill was lower at the time cool-season grasses exhibited their highest forage quality. Hughes and Harker (1950) found that weighing cattle 3 h after sunrise resulted in the lowest day-to-day variation in weight, with the exception of weighing after a 16-h fast. Similar results were observed by Taylor (1954) with grazing cattle. Consequently, the influence of season on amount of fill may need to be considered when deciding whether to use full or fasted weights. Also, fasting of animals before weighing involves the need for holding pens and may extend handling animals over 2 separate days rather than 1 d for nonfasted weighing.

Fasting may have adverse effects on some animals. In studies with cattle (Balch et al., 1953; Thornton & Yates, 1968), water deprivation reduced dry matter intake. Bond et al. (1976) studied feeding and drinking behavior of cattle following deprivation of feed, water, or both for 12- to 48-h periods. Withholding feed, water, or both from cattle for up to 2 d caused only minimal changes in subsequent intake; serum concentrations of Na, K, Ca, and Mg did not change as a result of fasting (Rumsey & Bond, 1976). Hughes (1976) reported that withholding feed or water for 12 h caused sheep (*Ovis aries*) and cattle to lose 5 to 6% of their body weight; weight loss doubled with a 36-h fast. Fasting longer than 12 h for sheep and 72 h for cattle may result in loss of tissue. Therefore, he suggested that sheep not be faster longer than 12 h and cattle from 12 to 16 h.

Matches (1981) concluded that either shrunk or unshrunk weighing was satisfactory for measuring season-long gains of cattle. However, he recommended that shrunk weighing (withholding of pasture and water) was desirable for short-term experiments because it helped minimize short-term variations in animal weight.

Automatic Weighing

Advances in electronics and animal identification, such as a pulse-code modulation system (Street, 1979), have enabled researchers (Anderson et al., 1981) to develop equipment that identifies and weighs animals automatically in a computer-compatible format without disturbing normal behavior. Weighing scales may be placed in strategic locations where animals regularly pass to reach water. Sorting gates set to an animal's electronic identification can direct an animal to its proper pasture. Subdermal body temperature and watering behavior also may be monitored electronically (Anderson et al., 1981). Adams et al. (1987) used a modification of this procedure in which a scale was placed in each pasture with data transmitted to a central facility.

Anderson and Weeks, as cited by Hart (1987), reported that their automatic weighing system could detect in 8 d the smallest significant difference in cow weights detected by manual weighing every 28 d. However, the dependability of their equipment was low in that it failed to operate for 344 d of a 910-d period. Adams et al. (1987) reported that their automatic weighing equipment functioned without problems for 45 d, during which time temperatures reached −30°C with snow on the ground. Advantages of automatic weighing methods include increased precision, reduced labor, and capacity to weigh in remote areas, even under extreme variations in weather.

Bovine "Boots"

A telmetry system (Horn & Miller, 1979; Horn, 1981b) was developed to continuously monitor bovine weight in order to measure intake. Animals wore four boots containing electronic load cells to generate analog signals, which were summed, converted to digital format, and transmitted to a receiver. Though novel, this appraoch proved impractical because animals damaged boots while attempting to discard them (Hart, 1987).

MEASUREMENT OF COMPOSITION OF LIVE TISSUE GAIN AND/OR BODY CONDITION

Changes in nutritional status of the grazing animal as well as composition of tissue gain may be more reflective of meaningful treatment effects than measurement of live weight changes (Koch et al., 1979). Also, lean meat is the ultimate product of forage/livestock production. Measurement of composition of animal live tissue gain is more difficult to determine than weight gain. Several authors have summarized techniques of measuring body composition (Corbett, 1978; Horn, 1981a; Anderson, 1987) as it relates to grazing research. Emphasis has been placed on composition of live body gain or loss as it relates to production efficiency rather than nutritional status of the animal. However, assessment of change in nutritional status due to grazing treatment could explain a variety of responses including changes in lean, fat, and bone. From an economic standpoint, it is important to characterize the ability of various forage systems to produce lean, fat, and bone. Also, it is important to understand energetic efficiency and tissue production or loss in both growing and mature (reproducing) animals.

Chemical Composition

The fat-free empty body of animals of a given species has a comparatively constant composition of water, protein, and ash (Burton & Reid, 1969). Because fat contains very little water, measurement of total body water can be used as an estimate of fat, fat-free lean, ash, and caloric content. Corbett (1978) suggested that estimates of total body water will usually include water in the gastrointestinal tract and that most variation associated with this may be overcome by a 24 to 48-h fast. Estimating total body water usually is accomplished by injecting a dye or marker that becomes distributed throughout the body water. After allowing time for equilibrium, the concentration of the injected marker is measured. Various markers can be used: tritiated water space (Panaretto, 1963; Panaretto & Till, 1963), deuterium oxide dilution (Houseman et al., 1973), antipyrine water space (Garrett et al., 1959), urea water space (Preston & Koch, 1973), Evans blue dye dilution (Anderson et al., 1969), and potassium-42 (Fuller et al., 1971).

Ultrasonic and Linear Measurements

Considerable research with ultrasonic methodology has been directed toward estimation of compositional differences among carcasses (Stouffer, 1966). Kempster (1981) concluded that levels of precision achieved with ultrasound techniques in sheep were of limited usefulness. Also, with sheep (Leymaster et al., 1985), the prediction equation developed from ultrasonic measurements with a Scanogram (Ithaco, Inc., Ithaca, NY) provided insufficient precision to use as a research tool. Bailey et al. (1986) found that inclusion of ultrasonic scanning measurements along with other live linear measurements did not improve prediction of fat, and lean, or of rib, loin, round, and hindshank muscle.

Technology associated with the use of ultrasonics has improved. Miller et al. (1988) reported reliable prediction of beef carcass fat by measurement of ultrasound shoulder fat and ultrasound rump fat plus ultrasound ribeye area, or by these measurements plus ultrasound 12th-rib fat. These studies included use of calves, feeders, yearlings, finished steers, and cows.

Unfortunately, measurement of body composition in the mature cow by methods beyond those based on weight, linear measurements, condition score, or possibly ultrasonics, are comparatively intense and difficult to apply. Thompson et al. (1983) suggested that linear measurements were not superior to condition scores in estimating body composition.

Gresham et al. (1986) developed equations for predicting carcass gross energy, protein, fat, and bone of beef cows. Equations that used subjective measures such as condition and frame score for predicting energy and carcass tissue weights were slightly more variable in precision than those developed with objective variables such as live weight, ultrasound fat between the first and second ribs and between the 12th and 13th ribs (10 cm lateral to the body midline), and wither height. However, a reasonable degree of accuracy and precision was achieved with both. These types of measurements may also be used to predict puberty (Brooks et al., 1985) and assess body condition as it relates to rebreeding of beef cows (Rutter & Randel, 1984; Richards et al., 1986).

Body Condition Scores

Although a subjective measure, condition scores can be used to estimate body composition and to assess general nutritional status of the animal when adequate standards are established. They also can be applied to pregnant animals where live weight changes may not be representative of the animal's nutritional status, and they offer estimates of body energy reserves.

Anderson (1987) described 19 scoring systems or standards based on visual appraisal, external palpation, or both. Visual scores are usually based on the amount of subcutaneous fat around the base of the tail or over the hips, back, or ribs. Lowman et al. (1976) developed a procedure in which five areas of the body are palpated for fat cover. These include the spinous processes of the lumbar vertebrae, the lower rib cage, the hip bone, around the tail head, and the gluteal muscles of the thigh. Nelsen et al. (1985) found no differences between palpation and visual measurements and their relationship to body weight, height, and heart girth.

Condition score has been associated with rebreeding of beef cows (Rutter & Randel, 1984; Richards et al., 1986), and milk production in dairy cows (Frood & Croxton, 1978; Garnsworthy & Topps, 1982), but the association of condition score and milk production in beef cows is less certain (Williams et al., 1979). Acceptance of the use of a condition scoring system will depend on repeatability. Body condition scores are useful for comparisons within an experiment; but since they are subjective, condition scores at different locations may be less reliable.

MEASUREMENT OF REPRODUCTION

Many grazed forages are consumed by breeding animals whose primary purpose is to produce healthy, high-performing offspring. Reproductive efficiency may be influenced by quality and amount of forage. However, plant constituents such as estrogenic compounds may influence conception rates (Moule et al., 1963).

Desired measurements will often differ between primaparous and multiparous females (Casida, 1968). For primaparous animals, age at puberty or the occurrence of first estrus is very important. For multiparous cows, measurements might begin with calving rate followed by pregnancy rate, conception rate, and postpartum interval (interval from calving to first estrus; Wagner & Oxenreider, 1972). In intensive studies, measurements might also include frequency of serum luteinizing hormone (LH) peaks and LH peak amplitude (Rawlings et al., 1980; Leung et al., 1986), which are parameters of gonadotropin secretion that become more dynamic as time for ovulation approaches. Serum progesterone concentrations greater than 1.0 ng/mL are an indication of prior ovulation (Arije et al., 1974; Radford et al., 1978). Unfortunately, very large numbers of animals are required to detect influence of pasture treatments on reproduction, especially for binary measures such as pregnancy.

MEASUREMENT OF MILK PRODUCTION

Milk production of grazing lactating females greatly influences preweaning performance of offspring. Because most dairy cows in the USA are not grazed, emphasis has been placed on beef cows and sheep. Milk yield is a sensitive indicator of the nutritive value of the diet. For example, changes in diet are quickly reflected by changes in milk yield (Dunn et al., 1969;

Jeffrey, 1970; Storry, 1970; Stobbs & Brett, 1972). High positive correlations (0.81–0.88) have been observed between estimated total lactation milk yield and calf weight (Neville, 1962; Totusek et al., 1973).

Milk yield of beef cows has been estimated by various methods including hand-milking while the calf nurses (Gifford, 1953), weighing the calf before and after nursing (referred to as the weigh-suckle-weigh technique by Knapp and Black, 1941; Neville, 1962; Totusek et al., 1973; Williams et al., 1979; and Boggs et al., 1980), and machine milking after oxytocin injection (Schwulst et al., 1966). Totusek et al. (1973) found that estimates of milk yield by the weigh-suckle-weigh procedure were less variable and averaged 29% higher than hand-milking at all stages of lactation. Milking after oxytocin injection gave similar results to the weigh-suckle-weigh technique with sheep (Doney et al., 1979) and cattle (Le Du et al., 1979). However, Beal et al. (1988) found that estimates of milk production obtained by machine milking following oxytocin injection gave results that were more repeatable and more highly correlated with calf weaning weight than those obtained by weigh-suckle-weigh. Williams et al. (1979) compared 4-, 8-, and 16-h weigh-suckle-weigh separation intervals and found that 8-h separation gave the best estimate of milk production during early lactation because it resulted in (i) less measurement error, (ii) higher correlation with calf average daily gain, and (iii) less irritation to the cows.

The estimation of milk production has been associated with the birth weight and sex of calf (Melton et al., 1967), frequency of nursing (Owen, 1957), interval between tests (Lampkin & Lampkin, 1960), and handling stress (Owen, 1957). Including composition of milk (fat and total solids) may not improve correlations of milk yield with calf weight (Totusek et al., 1973).

Schwulst et al. (1966) found that increased frequency of milk yield estimations gave higher correlation with calf gain. Measurements of milk production were less correlated with preweaning gain when calves were less than 30 d old (Schwulst et al., 1966; Totusek et al., 1973). Totusek et al. (1973) suggested that two to four properly timed estimates of milk yield could provide a good indicator of total yield. Rank of methods used to estimate milk yield in terms of accuracy and correlation with calf weaning weight from least to most desirable are hand-milking, weigh-suckle-weigh, and machine milking following oxytocin injection (Totusek et al., 1973; Beal et al., 1988).

MEASUREMENT OF FORAGE INTAKE

Herbage intake may help explain differences in animal response among forage treatments (Greenhalgh, 1982). A number of authors discussed advantages and disadvantages of methods for measuring forage intake of grazing animals (Cordova, 1978; Meijs, 1981; Leaver, 1982), none of which were completely satisfactory. These methods usually measure either decrease of herbage from a pasture or amount of consumption by animals.

The simplest method of measuring forage consumed was that of Erizian (1932) (as cited by Greenhalgh, 1982), who weighed the animals before and after grazing. Even though Erizian (1932) corrected for water consumed, urine and feces excreted, and respiratory losses, the method was not accurate.

Most methods for measuring intake are based on the concept that if the quantity of dry matter (or any nutrient) excreted in the feces of a grazing animal can be measured, and if the digestibility of the dry matter (or any nutrient) is known, then intake can be calculated [Intake = dry matter excreted/(1 − digestibility of dry matter)]. Methods utilizing this concept include the total collection technique and the marker technique, which include both external (substances given to the animal) or internal (substances naturally occurring in herbage) markers. These techniques usually require that the researcher obtain representative samples of dietary material for laboratory analysis and that forage indigestibility be accurately and precisely estimated.

Total Fecal Collection

Total fecal collection, measured by the use of harnesses and collection bags, is the oldest method for determining dry matter excreted by grazing livestock and is commonly referred to as the conventional or standard method (Schneider et al., 1955). A high labor requirement (Kartchner, 1975), reduced animal performance (Corbett, 1960), and poor estimates of digestibility (Holechek et al., 1986) are disadvantages of estimating intake via total fecal collection. However, animals have been fecal-bagged for 50 d (Greenhalgh et al., 1960) and 150 d (Raymond et al., 1953) without adverse effects. Phar et al. (1971) found no difference in intake between bagged and normal confined steers. A big advantage to bagging is that collection is rapid, requires only simple laboratory analysis, and provides estimates of feces over short periods of time. Cordova et al. (1978) concluded that the total fecal collection technique may be the technique of choice in many situations in spite of its disadvantages.

External Markers of Fecal Excretion

The premise of most external markers to estimate feces production is that the animal is fed a known amount of indigestible marker, which is assumed to be excreted in a specific known pattern. Kotb and Luckey (1972) summarized essential criteria for use of effective markers. Markers should (i) be inert and nontoxic, (ii) be quantitatively recovered in the feces (should not be absorbed or retained in the digestive tract), (iii) have no appreciable bulk, (iv) mix completely with the food and distribute uniformly during digestion, (v) have no influence on alimentary secretion, digestion, absorption, or motility of the microflora in the gastrointestinal tract, and (vi) be easy to analyze and inexpensive.

Although a number of substances have been investigated as markers, chromium sesquioxide (Cr_2O_3) is the one most commonly used. Most Cr_2O_3 methods have used discrete doses on a specified schedule followed by grab-sampling of feces (Corbett, 1960; Langlands et al., 1963). These methods are very labor intensive, disruptive to the animals, and involve episodic (usually diurnal) Cr_2O_3 excretion patterns. More recently, continuously controlled release devices for continuous delivery of Cr_2O_3 have been used to measure feces output of sheep (Ellis et al., 1981) and cattle (Ellis et al., 1982). Extensive studies using controlled-release chromic oxide devices (Barlow et al., 1988) suggested that fecal output measured by controlled-release chromic oxide worked well for estimating relative differences in dry matter intake between groups of cattle. They also suggested that large numbers of animals were needed to provide reasonable precision. Further, if absolute estimates of intake were required, it would be necessary to have reliable estimates of organic matter digestibility, and there would be a need to determine when to sample feces to get a representative estimate of marker excretion and fecal organic matter output.

Other markers have been used to estimate digesta rate of passage, as well as intake. These include rare earth elements such as ytterbium (Yb) absorbed onto hay and chromium (Cr) mordanted to the neutral detergent fiber of hay (Coleman et al., 1984).

Generally, when labor is not limiting, it appears that the total collection of feces is the superior way to measure dry matter excretion of grazing animals. The external marker technique requires more laboratory analyses and there are more opportunities for bias. Among the external markers, chromic oxide–impregnated paper appears to be the form of choice; however, controlled release chromic oxide appears promising.

Internal Indicators of Digestibility

The term *internal marker* or *indicator* has been used to refer to those naturally occurring in forage. Techniques that use internal indicators to estimate digestibility of forage grazed by cattle and sheep were reviewed by Wallace and Van Dyne (1970). They are valid only if representative samples of forage consumed and feces excreted are obtained and if the indicator is not digested. Wallace and Van Dyne (1970) concluded that lignin may be digested, particularly in immature forages, and could not serve as an internal indicator. Other proposed internal markers include chromogens or plant pigments (Reid et al., 1950) and silica (McManus et al., 1967), but for various reasons they have not been satisfactory.

Various fecal-index techniques to measure intake and digestibility have been investigated, including fecal nitrogen (N) in particular. The fecal N index procedure is based on the premise that fecal N is primarily of body origin and that metabolic fecal N is excreted in proportion to the quantity of dry matter consumed or digested (Blaxter & Mitchell, 1948). Equations utilizing the relationship between N concentration of feces and digestibility of forage have been developed (Lancaster, 1954), which can be used to estimate digestibility. A limitation to using fecal N to estimate intake is that fecal dry matter output must be determined. The fecal N index method has given valid estimates of digestibility (Wallace & Van Dyne, 1970; Scales et al., 1974); however, it appears less reliable as an indicator of intake (Cordova et al., 1978).

REPEATED MEASUREMENTS

According to Gill and Hafs (1971), caution must be used in statistical analysis of repeated measurements on the same animals because repeated measures are correlated and can be a problem if the correlations are not uniform. Statistical procedures are available that permit sensitive comparison of treatments when using repeated measures (Allen et al., 1983; Gill, 1986, 1988).

REFERENCES

Adams, D.C., P.O. Currie, B.W. Knapp, T. Mauney, and D. Richardson. 1987. An automated range-animal data acquisition system. J. Range Manage. 40:256–258.

Allen, O.B., J.H. Burton, and J.D. Holt. 1983. Analysis of repeated measurements from animal experiments using polynomial regression. J. Anim. Sci. 57:765–770.

Anderson, D.M. 1987. Direct measures of the grazing animal's nutritional status. p. 40–57. *In* Monitoring Animal Performance and Production Symp. Proc., Boise, ID. 12 February. Society for Range Management, Denver, CO.

Anderson, D.M., I. McDonald, and F.W.H. Elsley. 1969. The estimation of plasma and red cell volume in pigs. J. Agric. Sci. 73:501–505.

Anderson, D.M., J.A. Landt, and P.H. Salazar. 1981. Electronic weighing, identification and subdermal body temperature sensing of range livestock. p. 373–382. *In* J.L. Wheeler and R.D. Mochrie (ed.) Forage evaluation: Concepts and techniques. Am. Forage and Grassl. Council, Lexington, KY.

Arije, G.R., J.N. Wiltbank, and M.L. Hopwood. 1974. Hormone levels in pre- and post-parturient beef cows. J. Anim. Sci. 39:338–347.

Bailey, C.M., J. Jensen, and B.B. Andersen. 1986. Ultrasonic scanning and body measurements for predicting composition and muscle distribution in young Holstein × Friesian bulls. J. Anim. Sci. 63:1337–1346.

Bailey, P.J., and A.H. Bishop. 1975. Liveweight change, grazing time of steers and effect of pasture height on liveweight change following periods of hand feeding. Aust. J. Exp. Agric. Anim. Husb. 15:440–445.

Baker, A.L., R.W. Phillips, and W.H. Black. 1947. The relative accuracy of one-day and three-day weaning weights of calves. J. Anim. Sci. 6:56–59.

Baker, G.A., and H.R. Guilbert. 1942. Non-randomness of variations in daily weights of cattle. J. Anim. Sci. 1:293–299.

Balch, C.C., D.A. Balch, V.W. Johnson, and J. Turner. 1953. Factors affecting the utilization of food by dairy cows. 7. The effect of limited water intake on the digestibility and rate of passage of hay. Br. J. Nutr. 7:212–224.

Barlow, R., K.J. Ellis, P.J. Williamson, P. Costigan, P.D. Stephenson, G. Rose, and P.T. Mears. 1988. Dry-matter intake of Hereford and first-cross cows measured by controlled release of chromic oxide on three pasture systems. J. Agric. Sci. 110:217–231.

Bass, J.J., and D.M. Duganzich. 1980. A note on the effect of starvation on the bovine alimentary tract and its contents. Anim. Prod. 31:111–113.

Beal, W.E., R.M. Akers, and D.R. Notter. 1988. Milk production in beef cows: Methods of measurement and relationship to cow and calf performance. J. Anim. Sci. 66 (Suppl. 1):454.

Bean, H.W. 1948. Single weight versus a three-day average weight for sheep. J. Anim. Sci. 7:50–54.

Blaxter, K.L., and H.H. Mitchell. 1948. The factorization of the protein requirements of ruminants and of the protein value of feeds with particular reference to the significance of the metabolic fecal nitrogen. J. Anim. Sci. 7:351–372.

Boggs, D.L., E.F. Smith, R.R. Schalles, B.E. Brent, L.R. Corah, and R.J. Pruitt. 1980. Effects of milk and forage intake on calf performance. J. Anim. Sci. 51:550–553.

Bond, J., T.S. Rumsey, and B.T. Weinland. 1976. Effect of deprivation and reintroduction of feed and water intake behavior of cattle. J. Anim. Sci. 43:873–878.

Brown, M.A., and S.S. Waller. 1986. The impact of experimental design on the application of grazing research—an exposition. J. Range Manage. 39:197–200.

Brooks, A.L., R.E. Morrow, and R.S. Youngquist. 1985. Body composition of beef heifers at puberty. Theriogenology 24:235-250.

Burton, J.H., and J.T. Reid. 1969. Interrelationships among energy input, body size, age and body composition of sheep. J. Nutr. 97:517-524.

Casida, L.E. 1968. Studies on the postpartum cow. Univ. of Wisconsin Res. Bull. 270.

Coleman, S.W., B.C. Evans, and G.W. Horn. 1984. Some factors influencing estimates of digesta turnover rate using markers. J. Anim. Sci. 58:979-986.

Cook, C.W. 1986. Can grazing systems be studied with statistical validity. p. 9-17. *In* Statistical Analyses and Modeling of Grazing Systems Symp. Proc., Kissimmee, FL. 11 February. Society for Range Management, Denver, CO.

Cook, C.W., and J. Stubbendieck (ed.). 1986. Experimental design. p. 251-276. *In* Range research: Basic problems and techniques. Society for Range Management, Denver, CO.

Corbett, J.L. 1960. Faecal-index techniques for estimating herbage consumption by grazing animals. p. 438-442. *In* Proc. 8th Int. Grassl. Congr., Reading, England. 11-21 July. Alden Press, Oxford, England.

Corbett, J.L. 1978. Measuring animal performance. p. 163-231. *In* L. 't Mannetje (ed.) Measurement of grassland vegetation and animal production. Commonwealth Bureau of Pastures and Field Crops, Hurley, Bull. 52. Commonw. Agric. Bureaux, Farnham Royal, Bucks, England.

Cordova, F.J., J.D. Wallace, and R.D. Peiper. 1978. Forage intake by grazing livestock: A review. J. Range Manage. 31:430-438.

Dembiczak, C.M., H.D. Eaton, G. Beall, and H.L. Lucas, Jr. 1957. Design and conduct of calf nutrition studies. One vs. two and three-day growth measurements. J. Dairy Sci. 40:1133-1151.

Doney, J.M., J.N. Peart, and W.F. Smith. 1979. A consideration of the techniques for estimation of milk yield by suckled sheep and a comparison of estimates obtained by two methods in relation to the effect of breed, level of production and stage of lactation. J. Agric. Sci. 92:123-132.

Dunn, T.G., J.E. Ingalls, D.R. Zimmerman, and J.N. Wiltbank. 1969. Reproductive performance of two-year-old Hereford and Angus heifers as influenced by pre- and post-calving energy intake. J. Anim. Sci. 29:719-726.

Dwyer, D. 1961. Activities and grazing preferences of cows with calves in Northern Osage County, Oklahoma. Oklahoma Agric. Exp. Stn. Bull. B-588.

Ehrenreich, J.H., and A.J. Bjugstad. 1966. Cattle grazing time is related to temperature and humidity. J. Range Manage. 19:141-142.

Ellis, K.J., R.H. Laby, and R.G. Burns. 1981. Continuous controlled release administration of chromic oxide to sheep. Proc. Nutr. Soc. Aust. 6:145.

Ellis, K.J., R.H. Laby, P. Costigan, K. Zirkler, and P.G. Choice. 1982. Continuous administration of chromic oxide to grazing cattle. Proc. Nutr. Soc. Aust. 7:177.

Erizian, E. 1932. A new method for estimation of the quantity of pasture eaten by cattle. Z. Zuecht. Reihe B:25:443-459.

Frood, M.J., and D. Croxton. 1978. The use of condition-scoring in dairy cows and its relationship with milk yield and live weight. Anim. Prod. 27:285-291.

Fuller, M.F., R.A. Houseman, and A. Cadenhead. 1971. The measurement of exchangeable potassium in living pigs and its relation to body composition. Br. J. Nutr. 26:203-214.

Garnsworthy, P.C., and J.H. Topps. 1982. The effects of body condition at calving, food intake and performance in early lactation on blood composition of dairy cows given complete diets. Anim. Prod. 35:121-125.

Garrett, W.N., J.H. Meyer, and G.P. Lofgreen. 1959. An evaluation of the antipyrine dilution technique for the determination of total body water in ruminants. J. Anim. Sci. 18:1116-1126.

Gifford, W. 1953. Records of performance tests for beef cattle in breeding herds: Milk production of dams and growth of calves. Arkansas Agric. Exp. Stn. Bull. 531.

Gill, J.L. 1986. Repeated measurement: Sensitive tests for experiments with few animals. J. Anim. Sci. 63:943-954.

Gill, J.L. 1988. Repeated measurement: Split-plot trend analysis versus analysis of first differences. Biometrics 44:289-297.

Gill, J.L., and H.D. Hafs. 1971. Analysis of repeated measurements of animals. J. Anim. Sci. 33:331-336.

Goodchild, A.V. 1985. Gut fill in cattle: Effect of pasture quality on fasting losses. Anim. Prod. 40:455-463.

Greenhalgh, J.F.D. 1982. An introduction to herbage intake measurements. p. 1-10. *In* J.D. Leaver (ed.) Herbage intake handbook. British Grassl. Society, Hurley, England.

Greenhalgh, J.F.D., J.L. Corbett, and I. McDonald. 1960. The indirect estimation of the digestibility of pasture herbage. J. Agric. Sic. 55:377-383.

Gresham, J.D., J.W. Holloway, W.T. Butts, Jr., and J.R. McCurley. 1986. Prediction of mature cow carcass composition from live animal measurements. J. Anim. Sci. 63:1041-1048.

Harris, L.E., C.W. Cook, and J.E. Butcher. 1959. Symposium on Forage Evaluation: V. Intake and digestibility techniques and supplemental feeding in range forage evaluation. Agron. J. 51:226-234.

Hart, R.H. 1987. Monitoring changes in animal weights. p. 37-39. *In* Monitoring Animal Performance and Production Symp. Proc., Boise, ID. 12 Feburary. Society for Range Management, Denver, CO.

Holechek, J.L., and M. Vavra. 1983. Fistula sample numbers required to determine cattle diets on forest and grassland range. J. Range Manage. 36:323-326.

Holechek, J.L., H. Wofford, D. Arthun, M.L. Galyean, and J.D. Wallace. 1986. Evaluation of total fecal collection for measuring cattle forage intake. J. Range Manage. 39:2-4.

Horn, F.P. 1981a. Basic animal performance criteria for range and humid pastures. p. 299-312. *In* J.L. Wheeler and R.D. Mochrie (ed.) Forage evaluation: Concepts and techniques. Am. Forage and Grassl. Council, Lexington, KY.

Horn, F.P. 1981b. Direct measurement of voluntary intake of grazing livestock by telemetry. p. 367-372. *In* J.L. Wheeler and R.D. Mochrie (ed.) Forage evaluation: Concepts and techniques. Am. Forage and Grassl. Council, Lexington, KY.

Horn, F.P., and G.E. Miller. 1979. Bovine boots—a new research tool. p. 44-46. *In* 1979 Animal science res. report: Beef and dairy cattle, swine, sheep, poultry and their products. Oklahoma State Univ. Agric. Exp. Stn. MP-104.

Houseman, R.A., I. McDonald, and K. Pennie. 1973. The measurement of total body water in living pigs by deuterium oxide dilution and its relation to body composition. Br. J. Nutr. 30:148-155.

Hughes, G.P., and K.W. Harker. 1950. The technique of weighing bullocks on summer grass. J. Agric. Sci. 40:403-409.

Hughes, J.G. 1976. Short-term variation in animal liveweight and reduction of its effect on weighing. Anim. Breed. Abstr. 44:111-118.

Jeffery, H.J. 1970. The length of change-over periods in change-over design with grazing cattle. Aust. J. Exp. Agric. Anim. Husb. 10:691-693.

Johnson, W.M., and W.A. Laycock. 1963. Kind, number, and selection of livestock for grazing studies, and animal measurements most suited for evaluating results. p. 137-142. *In* Range research methods. USDA Misc. Publ. 940. U.S. Gov. Print. Office, Washington, DC.

Kartchner, R.J. 1975. Forage intake and related performance criteria of spring and fall calving cow-calf pairs on summer range. Ph.D. diss. Oregon State Univ., Corvallis.

Kempster, A.J. 1981. The indirect evaluation of sheep carcass composition in breeding schemes, population studies and experiments. Livest. Prod. Sci. 8:263-271.

Knapp, B., and W.H. Black. 1941. Factors influencing the rate of gain of beef calves during the suckling period. J. Agric. Res. 63:249-254.

Koch, R.M., R.P. Kromann, and T.R. Wilson. 1979. Growth of body protein, fat, and skeleton in steers fed on three planes of nutrition. J. Nutr. 109:426-436.

Koch, R.M., E.W. Schleicher, and V.H. Arthaud. 1958. The accuracy of weights and gains of beef cattle. J. Anim. Sci. 17:604-611.

Kotb, A.B., and T.D. Luckey. 1972. Markers in nutrition. Nutr. Abstr. Rev. 42:813-845.

Lampkin, K., and G.H. Lampkin. 1960. Studies on the production of beef from Zebu cattle in East Africa. II. Milk production in suckled cows and its effect on calf growth. J. Agric. Sci. 55:233-239.

Lancaster, R.J. 1954. Measurement of feed intake of grazing cattle and sheep. V. Estimation of the feed-to-feces ratio from the nitrogen content of the feces of pasture fed cattle. N.Z. J. Sci. Technol. Sect. A:36:15-20.

Langlands, J.P., J.L. Corbett, I. McDonald, and G.W. Reid. 1963. Estimation of the faeces output of grazing animals from the concentration of chromium sesquioxide in a sample of faeces. I. Comparison of estimates from samples taken at fixed times of day with faeces outputs measured directly. Br. J. Nutr. 17:211-218.

Leaver, J.D. (ed.). 1982. Herbage intake handbook. British Grassl. Society, Hurley, England.

Le Du, Y.L.P., J. Combellas, J. Hodgson, and R.D. Baker. 1979. Herbage intake and milk production by grazing dairy cows. 2. The effects of level of winter feeding and daily herbage allowance. Grass Forage Sci. 34:249-260.

Leung, K., V. Padmanabdon, L.J. Spicer, H.A. Tucker, R.E. Short, and E.M. Convey. 1986. Relationship between pituitary GnRH-binding sites and pituitary release on gonadotropins in postpartum beef cows. J. Reprod. Fertil. 76:53-63.

Leymaster, K.A., H.J. Mersmann, and T.G. Jenkins. 1985. Prediction of the chemical composition of sheep by use of ultrasound. J. Anim. Sci. 61:165-172.

Lowman, B.G., N. Scott, and S. Somerville. 1976. Condition scoring of cattle. East of Scotland College of Agric. Bull. 6. The Edinburgh School of Agric., Edinburgh, U.K.

Lush, J.L., and W.H. Black. 1927. How much accuracy is gained by weighing cattle three days instead of one at the beginning and end of feeding experiments. Am. Soc. Anim. Prod. Rec. Proc. Annu. Meet. 20:206-210.

Matches, A.G. 1970. Pasture research methods. Section I. p. 1-32. *In* R.F Barnes et al. (ed.) Proc. Natl. Conf. on Forage Quality Evaluation and Utilization, Lincoln, NE. 3-4 Sept. 1969. Nebraska Center for Continuing Education, Lincoln, NE.

Matches, A.G. 1981. Fill versus shrunk weights to estimate gain of cattle. p. 357-365. *In* J.L. Wheeler and R.D. Mochrie (ed.) Forage evaluation: Concepts and techniques. Am. Forage and Grassl. Council, Lexington, KY.

Meijs, J.A.C. 1981. Herbage intake by grazing dairy cows. Agric. Res. Rep. 909. Center for Agricultural Publishing and Documentation, Wageningen, the Netherlands.

McManus, W.R., M.L. Dudzinski, and G.W. Arnold. 1967. Estimation of herbage intake from nitrogen, copper, magnesium and silicon concentrations of faeces. J. Agric. Sci. 69:263-268.

Melton, A.A., J.K. Riggs, L.A. Nelson, and T.C. Cartwright. 1967. Milk production, composition and calf gains of Angus, Charolais, and Hereford cows. J. Anim. Sci. 26:804-809.

Miller, M.F., H.R. Cross, J.F. Baker, and F.M. Byers. 1988. Evaluation of live and carcass techniques for predicting beef carcass composition. Meat Sci. 23:111-129.

Morely, F.H.W. 1978. Animal production studies on grassland. p. 103-162. *In* L. 't Mannetje (ed.) Measurement of grassland vegetation and animal production. Commonwealth Bureau of Pastures and Field Crops, Hurley, Bull. 52. Commonw. Agric. Bureaux, Farnham Royal, Bucks, England.

Mott, G.O. 1959. Symposium on forage evaluation. IV. Animal variation and measurement of forage quality. Agron. J. 51:223-226.

Mott, G.O., and H.L. Lucas. 1953. The design, conduct, and interpretation of grazing trials on cultivated and improved pastures. p. 1380-1385. *In* Proc. 6th Int. Grassl. Congr., State College, PA. 17-23 Aug. 1952. Pennsylvania State Univ., State College, PA.

Moule, G.R., A.W.H. Braden, and D.R. Lamond. 1963. The significance of oestrogens in pasture plants in relation to animal production. Anim. Breed. Abstr. 31:139-157.

Nelsen, T.C., R.E. Short, W.L. Reynolds, and J.J. Urick. 1985. Palpated and visually assigned condition scores compared with weight, height and heart girth in Hereford and crossbred cows. J. Anim. Sci. 60:363-368.

Neville, W.E., Jr. 1962. Influence of dam's milk production and other factors on 120- and 240-day weight of Hereford calves. J. Anim. Sci. 21:315-320.

Obioha, F.C., D.C. Clanton, L.R. Rittenhouse, and C.L. Streeter. 1972. Source of variation in chemical composition of forage ingested by esophageal fistulated cattle. J. Range Manage. 23:133-136.

Owen, J.B. 1957. A study of the lactation and growth of hill sheep in their native environment and under lowland conditions. J. Agric. Sci. 48:387-411.

Panaretto, B.A. 1963. Body composition in vivo. II. The estimate of total body water with antipyrine and the relation of total body water to body fat in rabbits. Aust. J. Agric. Res. 14:594-601.

Panaretto, B.A., and A.R. Till. 1963. Body composition in vivo. V. The use of antipyrine tritiated water and *N*-acetyl-4-amino-antipyrine spaces. Aust. J. Agric. Res. 14:926-943.

Patterson, R.E. 1947. The comparative efficiency of single versus three-day weights of steers. J. Anim. Sci. 6:237-246.

Petersen, R.G., and H.L. Lucas. 1960. Experimental errors in grazing trials. p. 747-750. *In* Proc. 8th Int. Grassl. Congr., Reading, England. 11-21 July. Alden Press, Oxford, England.

Phar, P.A., N.W. Bradley, C.O. Little, L.V. Cundiff, and J.A. Boling. 1971. Nutrient digestibility using fecal collection apparatus and indicator methods for steers fed ad libitum. J. Anim. Sci. 33:695-697.

Preston, R.L., and S.W. Koch. 1973. In vivo prediction of body composition of cattle from urea space measurements. Proc. Soc. Exp. Biol. Med. 143:1057–1061.

Radford, H.M., C.D. Nancarrow, and P.E. Mattner. 1978. Ovarian function in suckling and non-suckled beef cows postpartum. J. Reprod. Fertil. 54:49–56.

Rawlings, N.C., L. Weir, B. Todd, J. Manns, and J.H. Hyland. 1980. Some endocrine changes associated with the postpartum period of the suckling beef cow. J. Reprod. Fertil. 60:301–308.

Raymond, W.F., C.E. Harris, and V.G. Harker. 1953. Studies on the digestibility of herbage. I. Technique of measurement of digestibility and some observations on factors affecting the accuracy of digestibility data. J. Br. Grassl. Soc. 8:301–314.

Reid, J.T., P.G. Woolfolk, C.R. Richards, K.W. Kaufman, J.K. Loosli, K.L. Turk, J.I. Miller, and R.E. Blaser. 1950. A new indicator method for determination of digestibility and consumption of forage by ruminants. J. Dairy Sci. 33:60–71.

Richards, M.W., J.C. Spitzer, and M.B. Warner. 1986. Effect of varying levels of postpartum nutrition and body condition at calving on subsequent reproductive performance in beef cattle. J. Anim. Sci. 62:300–306.

Rumsey, T.S., and J. Bond. 1976. Cardiorespiratory patterns, rectal temperature, serum electrolytes and packed cell volume in beef cattle deprived of food and water. J. Anim. Sci. 42:1227–1238.

Rutter, L.M., and R.D. Randel. 1984. Postpartum nutrient intake and body condition: Effect on pituitary function and onset of estrus in cattle. J. Anim. Sci. 58:265–274.

Scales, G.H., C.L. Streeter, A.H. Denham, and G.M. Ward. 1974. A comparison of indirect methods of predicting in vivo digestibility of grazed forage. J. Anim. Sci. 38:192–198.

Schneider, B.H., B.K. Soni, and W.E. Ham. 1955. Methods for determining consumption and digestibility of pasture forages. Washington Agric. Exp. Stn. Tech. Bull. 16.

Schwulst, F.J., L.J. Sumption, L.A. Swiger, and V.H. Arthaud. 1966. Use of oxytocin for estimating milk production of beef cows. J. Anim. Sci. 25:1045–1047.

Stobbs, T.H., and D.J. Brett. 1972. The effect upon the fatty acid composition of milk of feeding tropical pastures to Jersey cows. Proc. Aust. Soc. Anim. Prod. 9:297–302.

Storry, J.E. 1970. Reviews of the progress of dairy science: Ruminant metabolism in relation to the synthesis and secretion of milk fat. J. Dairy Res. 37:139–164.

Stouffer, J.R. 1966. Objective technical methods for determining carcass value in live animals with special emphasis on ultrasonics. World Rev. Anim. Prod. 1:59–66.

Street, M.J. 1979. A pulse-code modulation system for automatic animal identification. J. Agric. Eng. Res. 24:249–258.

Taylor, J.E. 1954. Technique of weighing the grazing animal. Proc. Br. Soc. Anim. Prod. p. 3–16.

Thompson, W.R., D.H. Theuninck, J.C. Meiske, R.D. Goodrich, J.R. Rust, and F.M. Byers. 1983. Linear measurements and visual appraisal as estimators of percentage empty body fat of beef cows. J. Anim. Sci. 56:755–760.

Thornton, R.F., and N.G. Yates. 1968. Some effects of water restriction on apparent digestibility and water excretion of cattle. Aust. J. Agric. Res. 19:665–672.

Totusek, R., D.W. Arnett, G.L. Holland, and J.V. Whiteman. 1973. Relation of estimation method, sampling interval and milk composition to milk yield of beef cows and calf gain. J. Anim. Sci. 37:153–158.

Wagner, W.C., and S.L. Oxenreider. 1972. Adrenal function in the cow. Diurnal changes and the effects of lactation and neurohypophyseal hormones. J. Anim. Sci. 34:630–635.

Wallace, J.D., and G.M. Van Dyne. 1970. Precision of indirect methods for estimating digestibility of forage consumed by grazing cattle. J. Range Manage. 23:424–430.

Whiteman, J.V., P.F. Loggins, D. Chambers, L.S. Pope, and D.F. Stephens. 1954. Some sources of error in weighing steers off grass. J. Anim. Sci. 13:832–842.

Williams, J.H., D.C. Anderson, and D.D. Kress. 1979. Milk production in Hereford cattle. II. Physical measurements: Repeatabilities and relationships with milk production. J. Anim. Sci. 49:1443–1448.

4 Measurements of the Plant-Animal Interface in Grazing Research[1]

S. W. Coleman
USDA-ARS
Forage and Livestock Research Laboratory
El Reno, Oklahoma

T. D. A. Forbes
Texas A&M University
Uvalde, Texas

J. W. Stuth
Texas A&M University
College Station, Texas

ABSTRACT

Grazed ecosystems are characterized by a dynamic, hierarchical interaction of soil, plants, and animals. Because of continuous flux between components and the impact of each component on the other components, the plant-animal interface has largely been neglected throughout many years of research. More conventional approaches have included studying one of the components while either ignoring the others or attempting to hold them static. Also, conventional experimental approaches to evaluation of grazing lands tend to measure responses over long periods of time, at infrequent intervals, seldom less than a grazing season of 12 to 16 weeks. Typical responses measured are net production, usually weight gain, or aboveground plant biomass. Gain is influenced by many factors, both of animal and plant origin, most notably dry matter intake and digestibility. Measurements of grazing behavior, in response to changes in the sward structure, chemical characteristics, and availability, lend themselves to short-term assessment of animal response compensations when foraging is restricted. Short-term intake can be calculated from measures of grazing behavior, and inferences can be made concerning the grazing strategy of various kinds of grazers. These parameters aid in the development of ecological and production models of the grazing system.

---

[1] Joint contribution no. PA24243 of the USDA-ARS and Texas Agric. Exp. Stn.

Copyright © 1989 Crop Science Society of America and American Society of Agronomy, 677 S. Segoe Rd., Madison, WI 53711, USA. *Grazing Research: Design, Methodology, and Analysis*, CSSA Special Publication no. 16.

Much grazing research has been conducted throughout the world to determine the performance of livestock on various kinds of pastures and ranges. Because of the diversity of the soil-plant animal (SPA) complex, different combinations of inputs may yield similar output, because of compensation by certain components of the system. Thus, much research can only be marginally applied to other management or environmental situations. If the dynamic nature of the SPA complex is considered and short-term changes in each component are monitored frequently, then the data can potentially be extrapolated to other situations, rather than limited to each geographic location of interest.

The impact of each component in the SPA complex is vital to understanding the complete system; yet, to design experiments to cover all of the possibilities is virtually impossible. Mathematical modeling coupled with artificial intelligence appears to be a logical approach to speed up the evaluation process. But, the model must be biologically correct, which requires that the interactions be included in data used for model development.

RESEARCH APPROACH

Malechek et al. (1986) expanded on the ideas of Caldwell (1984) to describe the transition from empirical to mechanistic experimentation. Initially, large-scale empirical grazing studies were conducted, which included many facets of management, but often only animal weight gain was measured. With the development of the esophageal-fistual technique (Cook et al., 1958), efforts began to describe the botanical and nutritional characteristics of the diet selected by the grazing animal, but little progress was made toward more mechanistic research. Behavioral aspects of the grazing system have been studied for some time (Johnstone-Wallace & Kennedy, 1944), but more germane questions have been asked in grazing research only in the last 20 yr. This has been stimulated in part by concepts from ecology and animal psychology, which initiated a new way of thinking about plant-animal relationships (Malechek et al., 1986; Stuth et al., 1987). We will approach the subject of the plant-animal interface from two views depicted in Fig. 4-1. The first will be the hierarchical interaction of the animal with the landscape. The second will focus on the plant's response and defense to grazing. Though the soil interacts, it will not be covered in detail in this chapter.

Hodgson (1985) argued the case for control and manipulation of sward conditions rather than control of stocking rate per se in research and management and laid the foundation for a more complete understanding of the interrelationships between plants and animals. Sward state is a function of soil, climatic, and management inputs such as grazing periodicity and intensity, fertilizer inputs, and supplemental feeding. The use of stocking rate (or associated variables such as grazing pressure or herbage allowance) as the control in grazing research does not allow the determination of input-output relationships because the sward state continually changes over the grazing season. Stocking rate cannot be considered as a primary component of either pasture production or animal performance, but may be used to create changes

MEASUREMENTS OF THE PLANT-ANIMAL INTERFACE

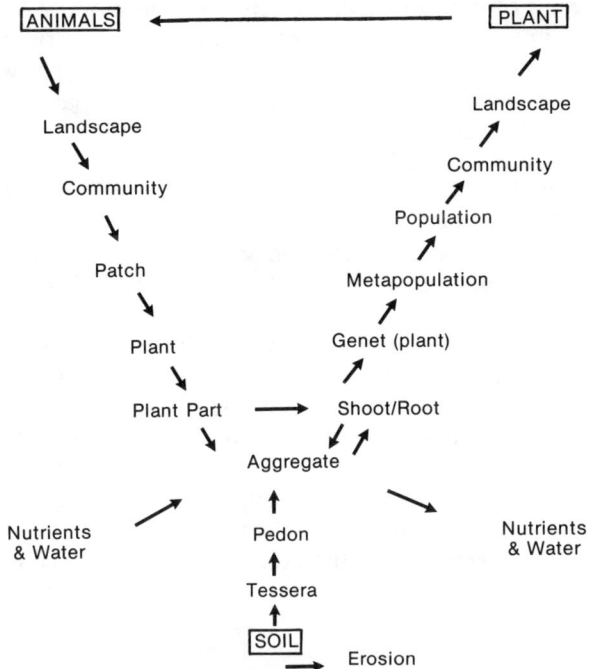

Fig. 4-1. The hierarchical context of the plant-animal interface.

in the sward for short duration (2-3 wk) experimentation. Although control of the sward may be the ideal research situation, especially with sown pastures, it may be argued that it is too difficult to achieve such control, especially with indigenous swards of warm-season species, due to rapid growth early in the season and dormancy during much of the year. However, within western rangelands of the USA, management of the sward was the obvious criterion for the "take half-leave half" philosophy (Hedrick, 1958).

HIERARCHY OF ANIMAL RESPONSE TO SWARD STRUCTURE

Drives Involved in Grazing and Selection

Ungulates graze to obtain necessary nutrients from the forage on offer. There are different drives that influence movement within a landscape, prioritized for domestic stock in the following order: (i) thirst or need for water balance; (ii) ambient temperature as it influences homeothermy; (iii) calorie balance or hunger, the drive which turns grazing on or off; (iv) time of day, especially nighttime as it influences orientation and predator avoidance; and (v) rumination, rest, social facilitation, and sleep (Stafford-Smith, 1988). Drives with the momentarily higher priority—for instance, the drive for water or the need to mitigate the effects of temperature extremes—may overrule the need for grazing. The intake drive for an individual animal is the net result of physiological nutrient demand modified by residues and metabolic end products of the last meal. In the grazing situation, the animal must, at a given time, not only choose whether to eat, but at which site (community), which level, which species, and which plant part. Grazing strategy of ungulates has become the subject of much study recently (Stuth et al., 1987; Malechek et al., 1986; Arnold, 1987).

Digestible dry matter or energy intake by grazing animals can be variable, due to heterogeneity of the sward, seasonal production, and variation in sward or canopy structure (Chacon & Stobbs, 1976). Ruminants have an enormous task of harvesting 100 to 400 g of fresh feed per kilogram of live weight daily. Spatial distribution of leaf within the sward or canopy influences the ease with which the animal can satisfy its need. With leafy temperate pastures, ruminants can consume forage in large bites and can satisfy their appetite rather easily in 6 to 8 h/d (Stobbs, 1973b). Time available for grazing is limited by rumination, need for social interaction, rest, and other factors. Cattle graze warm-season pastures for a longer time each day than temperate pastures, even when large quantities of herbage are available for grazing (Stobbs, 1974; Forbes & Coleman, 1987). Yet, rumination time is longer for warm-season forages than for temperate forages due to their greater fibrosity.

Foraging Tactics

The ungulate relates to the grazing situation in a hierarchical context (Senft et al., 1987; Fig. 4-2). The strategy for foraging is based on the need

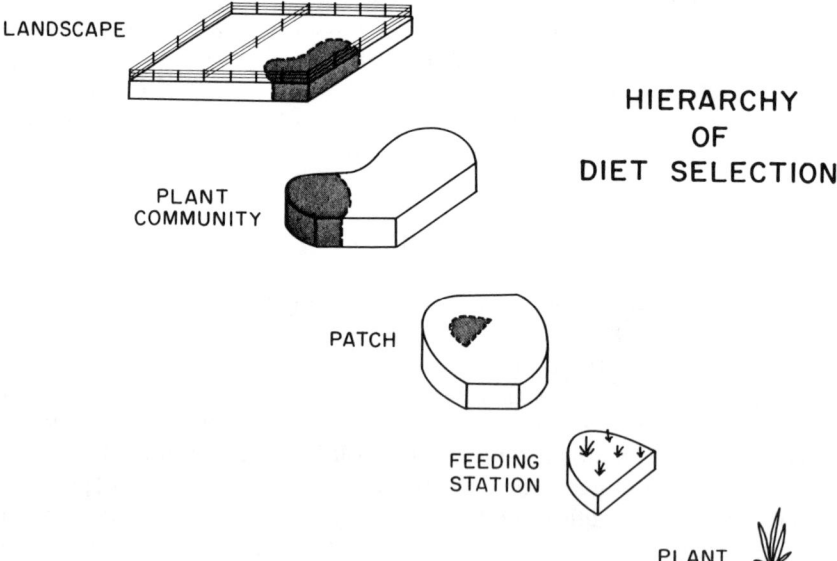

Fig. 4-2. The hierarchical context of diet selection as it descends from the landscape to the individual plant.

to maximize energy harvest rate at a feeding station and minimize energy expenditure between feeding stations. This is accomplished by matching the foraging behavior to the harvestibility of nutrients of a community encountered along a directional path as mediated by levels of the drives listed above, especially thirst, thermal balance, and satiety. Animal factors that influence foraging strategy do so predominantly through the drives noted previously. The drives may be expressed through classifications such as species, body size, and prehensile apparatus. There is a real need for new theory concerning foraging movement and decisions of ungulates.

Selection within the Landscape

The first decision the animal must make, after deciding to graze, is where. In sown pastures of mono- or di-cultures, choice is limited, since there is only one plant community. However, patch grazing, dung fouling, and location of water and mineral stations may alter the choice of certain patches over others. In extensive rangelands, spatial selection of plant communities and patches by the animal are influenced by the features of the landscape that affect animal movement patterns (Senft et al., 1987) including landscape boundaries; distribution of plant communities; accessibility and distribution of foci such as water, shade, and bedding sites; and other convergent points in a landscape.

Patchy grazing in sown pastures can cause nonuniform utilization of an otherwise uniform sward (Forbes & Coleman, 1987). When grazing pressure is too low, animals establish certain select locations in the pasture early in the season. If sufficient herbage is available, they may not graze outside the selected area for the remainder of the season. Under rangeland conditions, patch or mosaic grazing may be even more pronounced and have the devastating effect of eliminating vegetation in certain areas (Mott, 1987). Mosaic grazing is affected both by the grazer species and the characteristics of the sward. Patch grazing may cause one of the more difficult problems in conducting plant-animal interface research, because it is difficult to assess the forage available to the animal.

Species Selection

The feeding station is defined as an area that is grazed without taking a step. The animal chooses a grazing station, then initiates a search among species, plants, and plant parts. The animal must maximize the efficiency of use of time to satisfy the drive to fulfill nutrient requirements.

Within a feeding station, animals generally select diets that are higher in crude protein and digestibility than the average of that among the available forage. This positive selection pressure is a function of sward quality and amount of easily prehensible forage available. Stuth et al. (1987) found that species, previous defoliation, and number of live leaves were factors affecting the probability of an individual tiller being grazed among tall grass prairies in Texas. However, the influence of previous defoliation and number of leaves declined as stocking rate increased. Norton and Johnson (1983) found similar results with monotypic stands of crested wheatgrass [*Agropyron cristatum* (L.) Gaertn.]. Nearly 70% of all plants were grazed at least once, but only one-sixth of grazed plants were defoliated on more than one occasion. Observations of cattle grazing Asiatic bluestems in Oklahoma indicated that under light grazing, previously grazed areas were more likely to be regrazed than ungrazed areas, even before forage became stemmy with reproductive tissues (Forbes and Coleman, 1989, unpublished data). On moderately grazed swards, however, utilization was uniform across the pastures, similar to the patterns noted by Stuth et al. (1987) and Norton and Johnson (1983).

Diet Quality

Diet quality of grazing animals is usually higher than the average of that available in the sward because of the animals' ability to select certain sites, species, plants, and plant parts over less desirable ones (Coleman & Barth, 1974; Hodgson, 1982b). The more variable the sward, generally the greater the opportunity to select, but with greater investment in search time. Different classes and species of animals exhibit different degrees of selection (Hodgson, 1981b; Demment & Van Soest, 1981; Belovsky, 1984). Although it is generally accepted that animals select diets higher in quality than the average of the forage available, there is less agreement concerning the manner and mechanism for selection. Most researchers consider that animals can and do select certain parts of individual plants much as people select from a buffet table. However, Hodgson (1982b) found that percentage legume in the diet of cattle matched the percentage legume in the grazed horizon of a ryegrass-clover (*Lolium* sp.-*Trifolium* sp.) sward. This probably represents a sward density so great that even sheep could not distinguish among species, especially when the need for selection was low due to the excellent quality of both ryegrass and clover. Under tropical or subtropical conditions, this density of plant material most likely will not exist, but there is merit to the idea that cattle especially, which have a need for large volumes of bulk feed, may simply select patches and horizons with the patch. Once that decision is made, the material within the horizon is taken in the same proportion with which it occurs.

Ingestive Behavior

't Mannetje and Ebersohn (1980), in a review, noted that there was agreement on the general relationship between intake or animal performance and available herbage, but much variation exists in the slope and shape of the relationship. Simply measuring herbage mass may help explain why weight gains were different among treatments, stocking rates, or years, but the relationships were as limited for extrapolation as the conventional feed-and-weigh experiments. Causal relationships between intake or animal performance and pasture attributes can be defined only in terms fundamental to intake. Intake per day is a product of intake at each bite and the total number of bites taken in a day. The total number of bites is the product of biting rate and total grazing time. Each of these may vary depending on characteristics of the animal, the sward, or both.

Intake per Bite

Bite size may be defined as the amount of forage the animal prehends with one "head jerk" or severing motion. This is in contrast to jaw movements, which a number of researchers have used to define number of bites taken for a given amount of forage ingested (e.g., Stobbs & Cowper, 1972; Penning, 1983) and which include jaw movements to manipulate the forage for prehending or chewing after prehension. The ratio of manipulative jaw movements to bites tends to increase with increasing sward height (Chambers et al., 1981), increasing legume percentage (Moore et al., 1987), and/or intake per bite (Black & Kenney, 1984), making conversions unreliable. Therefore, it is concluded that total jaw movements are unreliable estimates of bites for calculating bite size. Bite size may be determined indirectly as a ratio of total bites per day to daily intake (Jamieson & Hodgson, 1979) or more directly by counting the bites taken by esophageally fistulated animals while the extrusa is being collected (Stobbs, 1973a; Forbes & Coleman, 1987).

Bite size is more responsive to sward conditions than are other parameters of ingestive behavior ('t Mannetje & Ebersohn, 1980; Forbes & Coleman, 1987; Forbes, 1988). As sward mass and height increase, bite size increases linearly in both tropical and temperate swards (Allden & Whittaker, 1970; Chacon & Stobbs, 1976; Hodgson, 1981a). However, Stobbs (1973b) found that bite size decreased with increasing maturity, and thus, height of tropical swards, but he attributed the relationship to low bulk density of leaf material in the upper grazing strata. However, he (Stobbs, 1973b) determined bite size by dividing the weight of extrusa collected by the total bites recorded by a jaw movement counter, which included manipulative as well as prehensive bites. Stemmy, less dense tropical swards with a high proportion of fiber would require more manipulative bites both before and after prehension, and thus, artificially reduce bite size.

Intake per bite is important to daily intake and is illustrated by the data of Hendrickson and Minson (1980), who noted that reducing bite size by 320 mg organic matter (OM) over a 12-d period resulted in a reduction in intake of almost 1 kg/d. Forbes (1988) summarized the literature concerning the relationship of bite size to sward height over a wide range of vegetation types and found considerable variability in the regression coefficients (0.02–0.09 mg OM/kg) liveweight per cm of plant height). However, the linear relationship is likely to decline into an asymptotic one once the sward becomes mature, seed stalks appear, and leaf density declines, and may be influenced by green leaf mass, dead leaf mass, green stem, and bulk density (Forbes & Coleman, 1989, unpublished data).

Grazing Time

The activity day of the animal is divided into periods of grazing, ruminating, and rest, the latter also including any social requirements (Hodgson,

1982a). The fraction of the day spent for each category depends on sward characteristics, climatic conditions, and grazing management. Grazing time interacts across all levels of hierarchy. At the landscape and plant community levels, it is predominated by search time, whereas at the feeding station level, ingestive behavior predominates. At the feeding station and plant levels, grazing time must be dissected into ingestion and search time. Often, measurements include all nonrest time, whereas in other instances only active grazing is included. The number of grazing periods and the duration of periods are somewhat dependent on the quality of the forage ingested in that poorer quality forage requires more rumination and resides longer in the rumen.

Estimates of time spent grazing may be obtained from continuous monitoring of activity or by using an interval sampling technique as described by Hodgson (1982a). Vibracorders (Stobbs, 1970) have been popular for determining both the total time and temporal patterns of grazing. They tend to measure total grazing activity, including search time if it is under 5 min, at a patch or even plant community level. Gary et al. (1970) found that recording activity at 15-min intervals provided reliable estimates for continuous activity of the animals.

Grazing time amounts to about one-third of the day for sheep (*Ovis aries*) and cattle (*Bos* sp.) (Forbes et al., 1985), but values have been found to range from 4.5 h (Hancock, 1954) to over 13 h/d (Stobbs, 1970) for cattle and sheep, depending on physiological needs of the animal and sward conditions. Environmental conditions may alter grazing patterns and total grazing time. If sward structure deteriorates to the extent that intake may be limited, the animal may increase the time searching and acquiring food. Poor sward structure often accompanies lower quality, which may increase the need for rumination, and thus, compound the problem of obtaining sufficient nutrients in the 24-h period.

In general there is a negative correlation between time spent grazing and sward height or herbage mass, but Chacon and Stobbs (1976) found poor correlations between grazing time and sward characteristics. The conditions of their experiment (a tropical grass sward grazed down over a 2-wk period) may have contributed to the poor results, as grazing time was negatively correlated to herbage mass or height during the 1st wk, but positively correlated the 2nd wk. Jamieson and Hodgson (1979) reported significant negative correlations between grazing time and green herbage mass for both sheep and cattle grazing temperate swards. Reports of trials with tall bunchgrass (*Bothriocloa* spp.) pastures in the USA indicated that grazing time declined as herbage mass, and thus, surface height of the sward increased (Forbes & Coleman, 1987). The function is probably asymptotic in shape, as are most biological systems, and reports of nonlinear relationships have occurred (Hendricksen & Minson, 1980). Method of grazing is clearly critical in experiments designed to improve our understanding of grazing behavior.

Rate of Biting

Rate of biting measurements have been conducted for many years as part of animal behavior (Johnstone-Wallace & Kennedy, 1944), but only recently have these been coupled with other parameters of behavior as fundamentals of intake. The method used to estimate biting rate is probably subject to more bias than other parameters of ingestive behavior. Because animals vary biting rate both within and between feeding stations, estimates should include measurements under both conditions (Stuth & Searcy, 1987). The 20-bite method used by Forbes and Hodgson (1985) was initiated while the animal was grazing within a feeding station. Although there may be little difference in rate of biting while the animal is searching within uniform sown swards under moderate to heavy continuous grazing, the rates between feeding stations under light grazing or on rangeland may be quite different. A good approach may include counting bites over a given time interval, probably 5 min, since that is the minimum resolution for estimation of a grazing period using vibracorders.

Rate of biting is quite variable among and within classes of livestock (Stobbs, 1974; Hodgson & Jamieson, 1981; Allden & Whittaker, 1970; Forbes et al., 1985) and ranges from 20 to 80 bites per minute for cattle and 18 to 120 for sheep. Forbes (1982) noted that cattle generally have faster rates of biting than sheep. Sheep are more selective to meet higher nutrient requirements, whereas cattle and other large ungulates are classified as grazing generalists.

In continuously grazed swards, rate of biting generally increases as sward height decreases, as a compensatory mechanism. Allden and Whittaker (1970), however, found that sheep on previously ungrazed plots increased biting rate as tiller length decreased from 35 to 5 cm and then decreased sharply as length was further decreased. Hodgson (1981a), with strip-grazed ryegrass, and Forbes and Coleman (1987), with continuously grazed Asiatic bluestem (*Bothriochloa* spp.), found little relationship of rate of biting under continuous grazing to sward height or mass. It is possible that sward characteristics were such that the animals were not required to compensate under these conditions. A more plausible explanation, especially for the conditions of Forbes and Coleman (1987), is that due to the structure of the sward at low herbage mass (< 1.2 Mg/ha), newly emerging leaves were protected by a prickly pad of very stiff stems, which prevented rapid biting rates. It was observed in these studies that steers spend considerable time using the tongue to manipulate green herbage into the mouth for severing while grazing on the swards of low herbage mass. This characteristic probably would not prevail with more lush temperate forages such as ryegrass, but may be prevalent with those similar in characteristics to the bluestems such as tall fescue (*Festuca arundinacea* Schreb.), orchardgrass (*Dactylis glomerata* L.), wheatgrass (*Agropyron* spp.), and other low density, tall growing bunchgrasses.

Intake per Day

Intake is one of the more important parameters in assessment of animal production in a management system, since rate of gain is highly dependent on digestible organic matter intake. Intake may be measured by indirect methods using markers (Hopper et al., 1978; Krysl et al., 1985) or by the product of different aspects of ingestive behavior. Forbes and Coleman (1989, unpublished data) observed that estimates of intake by ingestive behavior were higher than those using pulse dosed ytterbium (Yb) as a marker. This was the result of overestimating total bites per day, measured as the time required for the steers to take 20 bites, assuming the animal would continuously graze during a grazing period. This technique really measures maximum biting rate, a parameter that should change most with sward conditions. For estimating intake, however, average biting rate for a grazing period is required. Further, when using vibracorders, overestimates of grazing time may occur by erroneously ascribing short periods of activity (such as fighting or other social interactions) to grazing, but these errors are not as likely to occur as errors in biting rate measurement.

Intake measured by marker techniques can be adapted to describe the dynamic nature of intake across grazing season (Hopper et al., 1978). Though problems exist in all known methods of estimating intake of grazing animals, intake remains one of the most important measures of animal response because it is a fundamental of performance. Since intake per day is the product of bite size and total bites, factors that influence either of these fundamentals will influence intake. Often, management may affect the factors in such a way that intake may remain static over a range of sward conditions. Compensatory actions on the part of the animal to optimize nutrient intake is probably responsible for these effects; i.e., animals increase grazing time to compensate for reduced bite size.

HIERARCHY OF PLANT RESPONSE TO GRAZING

Grazing produces a complex situation where several dynamic processes interact with each other. Defoliation (Scott, 1956) and trampling (Edmond, 1963) by the animal influence plant growth and persistence. Plants with prostrate growth and high tiller population have more suitable characteristics for grazing (Hodgson, 1981a); however, erect plants have greater efficiency of incident light interception. Researchers have not fully considered the dynamics of growth and disappearance of herbage in the pasture ecosystem. Treatment effects are often confounded with the method used to measure production (Vickery, 1981). An almost continuous loss of pasture forage through death and decay is often overlooked.

Tiller or Individual Plant Tissue

Management of green leaf tissue should be the objective of the stockperson since leaves are the predominant structures for entrapment of solar energy. The dynamics of the plant portion of the pasture ecosystem are represented by leaves, tillers, and other plant parts that grow and die or are harvested (Davies, 1981). Rate of harvest by grazing will alter tiller (including both number and size) production and decline, leaf production, leaf senescence, and reproductive structure production. Grant et al. (1987) provided an excellent treatise on tissue flow and tiller dynamics for perennial ryegrass (*Lolium perenne* L.) swards. Tiller growth and loss and carbon dioxide flux were measured to determine the influence of grazing management on net canopy photosynthesis and its relationship to leaf area index (LAI).

Over a period of time, net change in the weight of living biomass is the result of the difference between gross gains in weight due to formation of new tissue and gross losses caused by the death, decomposition, and harvesting of older tissue. It is important to measure the gross as well as the net changes, because the scale of losses in grazed plots may be particularly high (Davies, 1981). This includes measuring leaf and stem growth rate, senescence rate, and harvest rate. In some instances, biomass may remain almost constant despite the continual growth and development of new tissue. The time lapse of conventional measurements of dry matter production is usually 1 to 2 wk or even longer (Parsons, 1981). The key to assessing the dynamics of the sward as affected by grazing is to shorten the measurement interval so that instantaneous rates may be assumed. Practically, these measurements may occur from 2 to 3 d to 1 wk as minimum intervals.

Grazing tolerance depends on replacing photosynthetic surface after it is removed (Archer & Tieszen, 1986). A critical part of research at the plant-animal interface is to create differences in the sward condition (structure) with different levels of grazing intensity, whether control is based on levels of stocking rate, herbage mass or height, or other variables. Grant et al. (1987) showed that interactions occurred between rotational or continuous grazing and level of intensity described by either LAI or sward height. Furthermore, net canopy photosynthesis was curvilinearly influenced by LAI during the regrowth phase and linearly influenced during the defoliation phase of rotational grazing. Therefore, timing of measurements in relation to the phase of the grazing cycle is critical, especially during rotational grazing.

Svejcar and Christiansen (1978b) found that heavy continuous grazing of Caucasian bluestem [*Bothriochloa caucasica* (Trin.) C. E. Hubb.] resulted in relatively larger reduction of LAI than of either root mass or length. This was accompanied by improved water status of leaf tissue in heavily grazed swards, resulting in reduced plant water stress as compared to leaves from lightly grazed swards (Svejcar & Christiansen, 1987a).

Nutrient Flows

Short-term assessment of sward state can be achieved by monitoring carbon exchange and nutrient or assimilate partitioning in the plant. Carbon exchange allows study of photosynthesis, whereas assimilate partitioning allows study of net production from photosynthesis. Although concentrations of total nonstructural carbohydrates (TNC) in continuously grazed Caucasian bluestem initially decreased more in heavily than in lightly grazed plants Christiansen and Svejcar (1987), the differences eventually disappeared as plants adjusted to grazing. There was little difference in amounts of TNC in stem bases regardless of grazing pressure. Part of the adaptation of the sward to grazing included compression and packing of stem bases in the crown of heavily grazed swards. This protection shielded the stem bases from grazing and maintained a pool of TNC available for regrowth and provided a safe location for new growing points to develop into tillers.

Plant Size and Density

As ungulates graze, removal of portions of tillers or plants elicits a response by the plants to replace the harvested matter. Butler and Briske (1988) and Olsen and Richards (1988) found that removing tillers from within a tussock, or removing neighboring tussocks from around a specific tussock were equally responsible for initiating new tillers. Research in Oklahoma indicated that heavier grazing pressure caused increased tiller number but smaller tillers in Caucasian bluestem (Christiansen & Svejcar, 1987).

Species Balance and Population Dynamics

Diversity

In indigenous pastures and ranges, where a great diversity of plant species and types occur, methods and objectives suitable for sown pastures may not be applicable. Under these circumstances, simple transfer of management practices and guidelines from one set of circumstances to another is not possible (Grant & Hodgson, 1986). Therefore, it is even more important to understand plant growth principles and survival and competition ecology among the plants. Behavior and selectivity of the grazer, discussed earlier, is a much greater factor under these circumstances.

The basic management objectives of indigenous pastures and ranges are to maintain a satisfactory species composition, maintain satisfactory levels of herbage production, and maintain adequate ground cover (Grant & Hodgson, 1986). The difficulties in understanding such swards is that several layers of organization are involved from the individual plant to the community (Fig. 4-1). Understanding dynamics at one level does not necessarily ensure understanding at higher levels, especially if measurements were made in isolated, protected, or controlled situations.

Archer and Tieszen (1986) observed that the survival, growth, and reproduction of a plant in a community depend on its ability to tolerate or avoid the stresses of the grazer. Generally plants have adapted a variety of mechanisms, which enable them to cope with abiotic conditions and grazing, and to compete with other species in the community.

Plant Defense

In order to survive, especially in indigenous communities, plants have developed defense mechanisms that were dealt with in detail by Malechek et al. (1986). Plants use chemical, morphological, and physiological measures to cope with grazing stress. Chemicals that repel or cause a negative selection pressure include tannins, oils, phenolics, and alkaloids. Morphological characteristics change with grazing, often offering some mode of physical protection to growing tips and young leaves. For instance, Caucasian bluestem has many small stem bases below a protective strata, which act to prevent grazers from destroying the crown. Yet, despite these defenses, some species may be eliminated from an ecosystem by overgrazing.

CONCLUSIONS AND RESEARCH NEEDS

In conclusion, measurements of the plant-animal interface are needed to understand the processes of the ecosystem. Short-term measurements of the dynamic process are necessary if mathematical models are to be developed that allow management alternatives to be compared. Understanding many of the processes is beginning to expand, but additional data, especially concerning the interactions of many of the inputs, are critically needed. Future needs at the landscape level for such models include (i) determination of decision rules involved in shifting and triggering needs; (ii) determination of grazing velocity and duration; and (iii) better knowledge of memory by animals.

The broad areas above can be broken down into critical knowledge needs. These include knowledge of (i) key community attributes that affect the relative profitability of a given combination of sites; (ii) the stimuli that animals use to begin and terminate a given activity; (iii) the stimuli that animals use to establish foci; (iv) the stimuli that animals use to set direction of travel to grazing sites, foci, or thermal niches; and (v) appropriate foraging behavior attributes that can be measured to assess foraging strategies of free-roaming animals.

REFERENCES

Allden, W.G., and I.A. McD. Whittaker. 1970. The determinants of herbage intake by grazing sheep: The interrelationship of factors influencing herbage intake and availability. Aust. J. Agric. Res. 21:755-766.

Archer, S.R., and L.L. Tieszen. 1986. Plant response to defoliation: Hierarchical considerations. p. 45-49. In O. Gudmundsson (ed.) Grazing research at northern latitudes. Plenum Press, New York.

Arnold, G.W. 1987. Influence of the biomass, botanical composition and sward height of annual pastures on foraging behaviour by sheep. J. Appl. Ecol. 24:759-772.

Belovsky, G.E. 1984. Moose and snowshoe hare competition and a mechanistic explanation from foraging theory. Oecologia 61:150-159.

Black, J.L., and P.A. Kenney. 1984. Factors affecting diet selection by sheep. II. Height and density of pasture. Aust. J. Agric. Res. 35:565-578.

Butler, J.L., and D.D. Briske. 1988. Population, structure and tiller demography of bunchgrass *Schizachyrium scoparium* in response to herbivory. Oikos 51:306-312.

Caldwell, M.M. 1984. Plant requirements for prudent grazing. p. 117-152. In Natl. Res. Council-Natl. Acad. Sci. Developing strategies for rangeland management. Westview Press, Boulder, CO.

Chacon, E.A., and T.H. Stobbs. 1976. Influence of progressive defoliation of a grass sward on the eating behaviour of cattle. Aust. J. Agric. Res. 27:709-727.

Chambers, A.R.M., J. Hodgson, and J.A. Milne. 1981. The development and use of equipment for the automatic recording of ingestive behaviour in sheep and cattle. Grass Forage Sci. 36:97-105.

Christiansen, S., and T. Svejcar. 1987. Grazing effects on the total nonstructural carbohydrate pools in caucasian bluestem. Agron. J. 79:761-764.

Coleman, S.W., and K.M. Barth. 1974. Quality of diets selected by grazing animals and its relation to quality of available forage and species composition of pastures. J. Anim. Sci. 36:687–692.

Cook, C.W., J.L. Thorne, J.T. Blake, and J. Edlefsen. 1958. Use of an esophageal-fistual cannula for collecting forage samples by grazing sheep. J. Anim. Sci. 17:189–193.

Davies, A. 1981. Tissue turnover in the sward. p. 179–208. In J. Hodgson et al. (ed.) Sward measurement handbook. The British Grassl. Society, Hurley, England.

Demment, M.W., and P.J. Van Soest. 1981. Body size, digestive capacity and feeding strategies of herbivores. Winrock Int. Livestock Res. Publ., Morrilton, AR.

Edmond, D.B. 1963. Effects of treading perennial ryegrass (*Lolium perenne* L.) and white clover (*Trifolium repens* L.) pastures in winter and summer at two soil moisture levels. N. Z. J. Agric. Res. 6:265–276.

Forbes, T.D.A. 1982. Ingestive behaviour and diet selection in grazing cattle and sheep. Ph.D. thesis. University of Edinburgh.

Forbes, T.D.A. 1988. Researching the plant-animal interface: The investigation of ingestive behavior in grazing animals. J. Anim. Sci. 66:2369–2379.

Forbes, T.D.A., and S.W. Coleman. 1987. Herbage intake and ingestive behavior of grazing cattle as influenced by variation in sward characteristics. p. 141–152. In F.P. Horn et al. (ed.) Proc. special session: Grazing-lands research at the plant-animal interface, Kyoto, Japan. 24–31 Aug. 1985. Winrock Int., Morrilton, AR.

Forbes, T.D.A., and J. Hodgson. 1985. Comparative studies of the influence of sward conditions on the ingestive behaviour of cows and sheep. Grass Forage Sci. 40:69–77.

Forbes, T.D.A., E.M. Smith, R.B. Razor, C.T. Dougherty, V.G. Allen, L.L. Erlinger, J.E. Moore, and F.M. Rougette, Jr. 1985. The plant-animal interface. p. 95–116. In V.H. Watson and C.M. Wells, Jr. (ed.) Simulation of forage and beef production in the southern region. Southern Coop. Ser. Bull. 308.

Gary, L.A., G.W. Sherrit, and E.B. Hale. 1970. Behavior of Charolais cattle on pasture. J. Anim. Sci. 30:203–206.

Grant, S.A., and J. Hodgson. 1986. Grazing effects on species balance and herbage production in indigenous plant communities. p. 69–77. In O. Gudmundsson (ed.) Grazing research at northern latitudes. Plenum Press, New York.

Grant, S.A., J. King, and G.T. Barthram. 1987. The role of sward adaptations in buffering herbage-production responses to grazing management. p. 21–32. In F.P. Horn et al. (ed.) Proc. special session: Grazing-lands research at the plant-animal interface, Kyoto, Japan. 24–31 Aug. 1985. Winrock Int., Morrilton, AR.

Hancock, J. 1954. Studies of grazing behavior in relation to grassland management. I. Variations in grazing habits of dairy cattle. J. Agric. Sci. 44:420–429.

Hedrick, D.W. 1958. Proper utilization—A problem in evaluating the physiological response of plants to grazing use: A review. J. Range Manage. 11:34–43.

Hendricksen, R. and D.J. Minson. 1980. The feed intake and grazing behaviour of cattle grazing a crop of *Lablab purpureus* cv. Rongai. J. Agric. Sci. 95:547–554.

Hodgson, J. 1981a. Variations in the surface characteristics of the sward and the short-term rate of herbage intake by calves and lambs. Grass Forage Sci. 36:49–57.

Hodgson, J. 1981b. Influence of sward characteristics on diet selection and herbage intake by the grazing animal. p. 153–166. In J.B. Hacker (ed.) Nutritional limits to animal production from pastures. Commonw. Agric. Bureaux, Farnham Royal, Bucks, England.

Hodgson, J. 1982a. Ingestive behaviour. p. 113–138. In J.D. Leaver (ed.) Herbage handbook. British Grassl. Society, Hurley, England.

Hodgson, J. 1982b. Influence of sward characteristics on diet selection and herbage intake by the grazing animal. p. 153–166. In J.B. Hacker (ed.) Nutritional limits to animal production from pasture. Commonw. Agric. Bureaux, Farnham Royale, Bucks, England.

Hodgson, J. 1985. The significance of sward characteristics in the management of temperate sown pastures. p. 63–67. In T. Okubo and M. Shiyomi (ed.) Proc. 15th Int. Grassl. Congr. Kyoto, Japan. 24–31 Aug. The Natl. Grassl. Res. Inst., Nishi-narino, Japan.

Hodgson, J., and W.S. Jamieson. 1981. Variations in herbage mass and digestibility, and the grazing behaviour and herbage intake of adult cattle and weaned calves. Grass Forage Sci. 36:39–48.

Hopper, J.T., J.W. Holloway, and W.T. Butts, Jr. 1978. Animal variation in chromium sesquioxide excretion patterns of grazing cows. J. Anim. Sci. 46:1096–1102.

Jamieson, W.S., and J. Hodgson. 1979. The effects of variation in sward characteristics upon the ingestive behaviour and herbage intake of calves and lambs under a continuous stocking management. Grass Forage Sci. 34:273–282.

Johnstone-Wallace, D.B., and K. Kennedy. 1944. Grazing management practices and their relationship to the behavior and grazing habits of cattle. J. Agric. Sci. 34:190-197.

Krysl, L.J., F.T. McCollum, and M.L. Galyean. 1985. Estimation of fecal output and particulate passage rate with a pulse dose of ytterbium-labeled forage. J. Range Manage. 38:180-182.

Malechek, J.C., D.F. Balph, and F.D. Provenza. 1986. Plant defense and herbivore learning: Their consequences for livestock grazing systems. p. 193-208. In O. Gudmundsson (ed.) Grazing research at northern latitudes. Plenum Press, New York.

't Mannetje, L., and J.P. Ebersohn. 1980. Relations between sward characteristics and animal production. Trop. Grassl. 14:273-280.

Moore, J.E., L.E. Sollenberger, G.A. Morantes, and P.T. Beede. 1987. Canopy structure of *Aeschynomene americana-Hemarthria altissima* pastures and ingestive behavior of cattle. p. 93-114. In F.P. Horn et al. (ed.) Proc. special session: Grazing-lands research at the plant-animal interface, Kyoto, Japan. 24-31. Aug. 1985. Winrock Int., Morrilton, AR.

Mott, J.J. 1987. Patch grazing and degradation in native pastures of the tropical savannas in northern Australia. p. 153-161. In F.P. Horn et al. (ed.) Proc. special session: Grazing-lands research at the plant-animal interface, Kyoto, Japan. 24-31. Aug. 1985. Winrock Int., Morrilton, AR.

Norton, B.E, and P.S. Johnson. 1983. Pattern of defoliation by cattle grazing crested wheatgrass pastures. p. 462-464. Proc. 14th Int. Grass. Congr., Lexington, KY. 15-24 June 1981. Westview Press, Boulder, CO.

Olsen, B.E., and J.H. Richards. 1988. Spatial arrangement of tiller replacement in *Agropyron desertorum* following grazing. Oecologia 76:7-10.

Parsons, A.J. 1981. Carbon exchange and assimilate partitioning. p. 209-227. In J. Hodgson et al. (ed.) Sward measurement handbook. British Grassl. Society, Hurley, England.

Penning, P.D. 1983. A technique to record automatically some aspects of grazing and ruminating behaviour in sheep. Grass Forage Sci. 38:89-96.

Scott, J.D. 1956. The study of perennial buds and the reaction of roots to defoliation as a basis of grassland management. p. 479. In G.J. Neale (ed.) Proc. 7th Int. Grassl. Congr., Palmerston North, New Zealand. Wright and Carmen Ltd., E. Wellington, New Zealand.

Senft, R.L., M.B. Coughenour, D.W. Bailey, L.R. Rittenhouse, O.D. Salla, and D.M. Swift. 1987. Large herbivore foraging and ecological hierarchies. BioScience. 37:789-799.

Stafford-Smith, M. 1988. Modeling: Three approaches to predicting how herbivore impact is distributed in rangeland. New Mexico Agric. Exp. Stn. Reg. Res. Rep. 628.

Stobbs, T.H. 1970. Automatic measurement of grazing time by dairy cows on tropical grass and legume pastures. Trop. Grassl. 4:237-244.

Stobbs, T.H. 1973a. The effect of plant structure on the intake of tropical pastures. I. Variation in the bite size of grazing cattle. Aust. J. Agric. Res. 24:809-819.

Stobbs, T.H. 1973b. The effect of plant structure on the intake of tropical pastures. II. Differences in sward structure, nutritive value and bite size of animals grazing *Setaria anceps* and *Chloris gayana* of various stages of growth. Aust. J. Agric. Res. 25:821-829.

Stobbs, T.H. 1974. Components of grazing behaviour of dairy cows on some tropical and temperate pastures. Proc. Aust. Soc. Anim. Prod. 10:299-302.

Stobbs, T.H., and L.J. Cowper. 1972. Automatic measurement of the jaw movements of dairy cows during grazing and rumination. Trop. Grassl. 6:107-112.

Stuth, J.W., J.R. Brown, P.D. Olson, M.R. Araujo, and H.D. Aljoe. 1987. Effects of stocking rate on critical plant-animal interactions in a rotationally grazed *Schizachyrium-Paspalum* savannah. p. 115-139. In F.P. Horn et al. (ed.) Proc. special session: Grazing-lands research at the plant-animal interface, Kyoto, Japan. 24-31 Aug. 1985. Winrock Int., Morrilton, AR.

Stuth, J.W., and S. Searcy. 1987. A new electronic approach to monitoring ingestive behavior of cattle. p. 81-82. In M. Rose (ed.) Proc. 2nd Int. Symp. Nutr. of Herbivores. St. Lucia, Queensland, Australia. 6-10 July. Aust. Soc. Anim. Prod., Brisbane, Australia.

Svejcar, T., and S. Christiansen. 1987a. Grazing effects on water relations of caucasian bluestem. J. Range Manage. 40:15-18.

Svejcar, T., and S. Christiansen. 1987b. The influence of grazing pressure on rooting dynamics of caucasian bluestem. J. Range Manage. 40:224-227.

Vickery, P.J. 1981. Pasture growth under grazing. p. 55-77. In F.H.W. Morley (ed.) World animal science. B1: Grazing animals. Elsevier Science Publ., New York.

5 Compromises in the Design and Conduct of Grazing Experiments

David I. Bransby

Auburn University
Auburn University, Alabama

ABSTRACT

Financial and logistical constraints usually restrict the number of animals and paddocks in grazing experiments. This forces a compromise in key elements of design, such as number of treatments, stocking rates (or levels of herbage mass if variable stocking is used), and replicates. Stocking procedures, such as stocking method (fixed vs. variable), are also important considerations in grazing experiments because they affect interpretation and application of results, and management requirements. Strengths and weaknesses of alternative designs and stocking procedures are discussed for grazing experiments with different objectives. Replication provides the best estimate of experimental error but restricts the number of treatments and/or stocking rates within constrained resources. Nonreplicated, multiple fixed, or variable stocking rate designs and partial replication constitute alternatives that may make more efficient use of restricted facilities than do traditional replicated designs, to meet specific objectives.

If production per animal and per hectare are response variables of interest in a grazing experiment, the experimental unit is the paddock *and* the animals that graze it. The discussion that follows considers only animal production experiments and not plant-animal interface experiments in which animal production is not measured. Because of the large size and cost of each experimental unit (including funding for land, animals, fencing, water lines and troughs, handling facilities, management, and labor), the number of units in grazing studies is usually restricted. This necessitates restrictions in the scope (number and nature) of the objectives, and in the number of treatments and replicates. Choice of the most appropriate compromise in design and stocking procedure depends on knowledge of available options, strengths and weaknesses of each, and efficiency of the approach relative to objec-

---

Copyright © 1989 Crop Science Society of America and American Society of Agronomy, 677 S. Segoe Rd., Madison, WI 53711, USA. *Grazing Research: Design, Methodology, and Analysis*, CSSA Special Publication no. 16.

tives. The following discussion is intended to assist researchers in examining alternative stocking procedures and designs for grazing experiments, and in making informed choices.

STOCKING CONSIDERATIONS IN GRAZING EXPERIMENTS

Stocking procedures must be considered in association with the design of grazing experiments because they strongly influence interpretation and application of data. The most important stocking considerations are stocking method (fixed vs. variable), number of grazing levels per treatment (single vs. multiple), and method of grazing management (continuous vs. rotational). However, the latter category will not be discussed. The term *grazing level* is used in a general context to refer to sward state, as indicated by one or more sward descriptors such as height or herbage mass per hectare (herbage mass). Different grazing levels can be obtained by grazing a treatment at several fixed stocking rates or herbage allowances/availabilities (via variable stocking). Therefore, grazing level can also be expressed in terms of animal numbers, as stocking rate or animal grazing days per hectare. *It is assumed that the value of using multiple grazing levels per treatment is mostly in increasing utility of absolute results rather than in testing differences among grazing levels. Consequently, grazing level itself will not be considered a treatment.*

Stocking Methods

Methods of stocking grazing experiments can be divided into the two broad categories of variable, or put-and-take (P&T) stocking, and fixed stocking. Variable stocking requires adjustment of animal numbers or land area to equalize plant defoliation and animal feeding conditions over time. In fixed stocking experiments, stocking rate (animals per ha) is kept constant and pasture sward conditions are allowed to vary across treatments and replicates over time (Wheeler, 1962).

The P&T method is well documented (Mott & Lucas, 1952; Mott, 1960; Petersen & Lucas, 1968; Matches, 1970). Wheeler et al. (1973) provided a detailed account of many factors that influence the choice between fixed and variable stocking. In general, fixed stocking can be used when (i) the pasture growth pattern is uniform and predictable; (ii) quality and quantity of forage can be maintained if forage is not consumed, as in many rangeland situations; (iii) flexibility in experimental conditions and related production conditions is limited; and (iv) results are directly applied to farm or range practice and used for economic analysis. Variable stocking is preferred when (i) the pasture growth pattern is markedly periodic and/or unpredictable; (ii) quality and quantity of forage cannot be maintained if forage is left unconsumed, as for many humid improved pasture situations; (iii) experimental conditions and related production conditions are flexible; and (iv) results are extrapolated to practice. Matches and Mott (1975) provide a good example of how such extrapolation can be achieved. An assessment of the factors above should help to indicate the most appropriate alternative.

Despite general agreement on the rationale above, several issues deserve discussion. Experimenters who use fixed stocking often employ several stocking rates per treatment, but may not recognize the value of estimating the effects of sward characteristics on animal production. This is apparent from common emphasis on the relationship between production per animal and stocking rate in these studies. For example, Jones (1981) suggested that inferences could be drawn about pasture treatments from the nature of this relationship. Although this may be partly true, without sward data interpretation is restricted, because the production per animal vs. stocking rate relationship simply relates production per animal to animals per hectare. Therefore, fixed stocking experiments have often been weak on interpretation. If the production per animal vs. herbage mass, and stocking rate vs. herbage mass relationships are examined, interpretation is more meaningful (Bransby, 1985; Bransby et al., 1988).

Originally, Mott and Lucas (1952) suggested that output per animal was a measure of forage quality (specifically a function of nutritive value and rate of intake). They indicated that nutritive value and palatability of forages are greatly influenced by physiological condition of the sward. Therefore, physiological condition and consumption of forage should be kept optimal, or alternatively, objective records of physiological condition and forage available per animal should be kept. Subsequently, Mott (1960) described the P&T procedure more rigorously. Having defined grazing pressure as the number of animals per unit of available forage, he emphasized that "it is imperative that the grazing pressure imposed on each of the treatments should be equal" and that "testers measure the quality of the herbage." Matches (1970), citing several others, stated that "Proponents of the put-and-take method contend that representative comparisons between pasture treatments are obtained only when grazing pressure is maintained at uniform levels within all treatments and across all treatments."

From comparison of the earlier literature with more recent work, it is apparent that several changes in the P&T procedure have occurred. Firstly, some workers (e.g., Matches et al., 1981) use grazing pressure (as originally described by Mott, 1960) as a basis for adjusting animal numbers or land area, but others use alternative sward descriptors such as herbage mass, sward height, mass of leaf material per hectare, etc. (e.g., McLaren et al., 1983; Burns et al., 1984; Sollenberger et al., 1988). In general, use of grazing pressure seems to have been restricted mainly to rotational grazing, whereas the other descriptors are more commonly associated with continuous grazing, but can be used with either grazing method.

Secondly, use of production per animal to measure forage quality, as suggested by Mott and Lucas (1952), Mott (1960), and Petersen and Lucas (1960), assumes (i) that only the chosen sward descriptor and forage quality influence differential output per animal and (ii) that sward descriptor is not severely confounded with treatment. Clearly, each of these conditions will apply only to a certain degree in each experiment. Furthermore, unintentional confounding of sward descriptor with replicate can inflate the error term. Levels of such imperfections that are scientifically acceptable are therefore subject to judgement, and may vary widely among researchers. For example, some attempts at equalizing sward height and/or herbage mass have resulted in large statistically significant differences from 500 to 1300 kg ha^{-1} within and/or across treatments and replicates over time (Burns et al., 1984; Read & Camp, 1986; Sollenberger et al., 1988). Some researchers may consider this degree of control adequate. However, differences of this magnitude will often translate into relatively large differences in output per animal, weakening statistical tests and restricting interpretation. Furthermore, if treatments are defined in terms of herbage mass, and replicates are paddocks treated alike, severe unintentional confounding of herbage mass with replicate makes the existence of formal replication questionable.

Finally, herbage mass and treatment are sometimes confounded deliberately on the grounds that treatments differ in their tolerance to defoliation intensity (e.g., McLaren et al., 1983; Burns et al., 1983; Sollenberger et al., 1988). Although this may be justified, it precludes separation of herbage mass (or any other sward descriptor) effects from other treatment effects, thus deviating from the original P&T guidelines.

In summary, over the past several years the P&T concept has undergone considerable divergent evolution. This process has resulted in the advantages of increased flexibility and utility of the procedure. However, researchers who consider using P&T should be aware that, depending on conditions and how it is used, control of pasture canopy characteristics may be no better than under fixed stocking. Furthermore, results may differ very little between the two methods, provided several appropriate levels of grazing are employed per treatment (Burns et al., 1970; Marten & Jordan, 1972). Despite this, if available resources do not permit more than one grazing level per treatment, P&T would probably be the method of choice, because it offers greater opportunity to graze at an estimated optimum level of herbage mass. If each treatment can be grazed at several levels, the choice appears to be less critical.

Grazing Level

Mott and Lucas (1952) made the following observation:

> In most grazing experiments, the stocking rate is adjusted to the level which is thought to be optimum for a given treatment. Imperfect knowledge of the [location of the] optimum for various treatments as well as failure to attain the optima may result in biased comparisons. Hence, in many types of work it may be necessary to test all treatments at three or more stocking levels. This allows the optimum stocking rate to be ascertained for each treatment, and the several treatments to be compared at their optimum.

Burns et al. (1970), Matches (1970), and others have made similar statements. Yet, Wheeler et al. (1973) point out that, apart from two methodological studies (Burns et al., 1970; Marten & Jordan, 1972), few P&T experiments have used three or more grazing levels per treatment. This may be related to the perceived need to replicate to provide a valid error term, and to financial constraints that restrict the number of paddocks in most grazing experiments. However, procedures that allow statistical analysis without formal replication (to be discussed later) are available.

In addition to the benefits identified by Mott and Lucas (1952), use of multiple grazing levels per treatment in grazing experiments offers the following advantages:

1. Treatment × grazing level interactions, which are quite common (Riewe, 1984), can be detected. If interactions of this nature exist such that response lines intersect (e.g., Bransby, 1988; Bransby et al., 1988) but each treatment is grazed at only one level, results will have very narrow application. In addition, they can be misleading, particularly for the unsuspecting producer. For example, if single grazing level experiments were conducted at low, medium, or high stocking rates for the situation presented in Fig. 5-1, average daily gain for Treatment A would be higher, the same as, and lower than for Treatment B at low, medium, and high stocking rates, respectively. All outcomes are therefore possible, depending on the stocking rate chosen. Results can therefore be misleading, because extension personnel and producers cannot be expected to anticipate this situation with access to data from only one grazing level.

2. Multiple grazing levels can ensure that a certain range in herbage mass (and other sward descriptors) will be common to all treatments. By comparing production per animal vs. herbage mass relationships for different treatments within this range, effects of herbage mass can be separated from treatment effects. This can also be done for other sward descriptors.

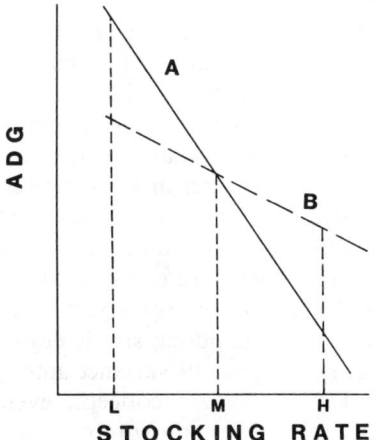

Fig. 5-1. Possible average daily gain (ADG) responses for hypothetical Treatments A and B showing a treatment × stocking rate interaction such that all outcomes are possible with a single grazing level experiment, depending on whether a low (L), medium (M), or high (H) stocking rate was chosen.

3. Good stand longevity for perennial species is a trait that is highly valued among producers. Under production conditions there are often times when managers find it difficult to avoid heavy grazing. Multiple grazing levels can be helpful in determining the approximate tolerance of pasture species to close grazing, thus providing an indication of the practical consequences of such situations.

4. Animal production responses to grazing level are particularly useful for economic analysis, because the economic optimum grazing level can be estimated even though it may change with treatment, market conditions, and other factors.

5. Effects of poor control of pasture canopy conditions are less critical when multiple grazing levels (as opposed to a single level) are used, because they can be largely removed in the multiple level analysis.

6. Data provided by multiple grazing levels are well-suited to modeling.

Assignment of multiple grazing levels per treatment in a grazing study requires several practical and analytical considerations. First, it is useful to have an appropriate range in sward condition common to all treatments. In the absence of dry matter yield data and previous grazing records, P&T stocking is likely to more effectively achieve this goal. If fixed stocking is to be used, yield data from clipping studies may help to determine a suitable range of stocking rates for each treatment. Failure to achieve both heavy and light grazing on each treatment could lead to bias. It does not matter that these stocking rates are often outside the range of economic importance, because extremes are critical in maximizing precision of resultant response functions. Neither should there be undue concern about the heavy stocking rate being so high that the pasture is damaged (provided it is not killed prematurely), because this can assist in providing a rough indication of the grazing level beyond which stand loss occurs. In such a case it is particularly useful to have more than three grazing levels per treatment.

A second consideration is how to create different grazing levels. The options are keeping land area constant and varying animal numbers, keeping animal numbers constant and varying land area, or varying both. Confounding is clearly inevitable. In general, varying animal numbers on equal land area results in heterogeneous variance among animals within pastures, whereas variation in land area will result in heterogeneous variance among paddocks. Some statistical implications of each strategy are discussed by Petersen and Lucas (1960). Varying land area usually minimizes land and animal requirements, and is therefore cost-effective, but will make rerandomization difficult if the site is used for several experiments in sequence. From this point of view, equal paddock size is desirable. With equal paddock size the problem of heterogeneous variance among animals can be overcome by using the tester and grazer concept, even if fixed stocking is employed. An equal number of testers from each paddock can be used to calculate average production per animal, while grazers are used to create the different grazing levels.

Finally, criteria for initiating and terminating grazing treatments and grazing levels deserve discussion. Grazing can be initiated on all paddocks at the same time, or at different times, when the sward descriptor reaches a predetermined goal for each paddock. Similarly, grazing of all paddocks could be terminated simultaneously, or at different times, depending on sward descriptor goals or predetermined levels of production per animal. If grazing is initiated and terminated at different times, this should be done on an objective basis. Length of the grazing period can then be treated as a response variable. However, it will be confounded with treatment and grazing level, and careful consideration is needed for data interpretation and application. Because these issues are independent of experimental design and procedure, consideration should be given to analyzing and reporting results over time in all grazing experiments.

KEY ELEMENTS OF DESIGN

In designing grazing experiments it is necessary to consider biological, practical, and statistical implications. Although these three categories are interrelated, this breakout permits identification of certain key elements of design that relate to each.

Biological Considerations

The biological value of a grazing experiment is the extent to which the study adds to existing biological information on, and understanding of, grazing ecosystems. This biological value is increased when the number of treatment comparisons is increased, making treatment number a key element of design. For example, a two-treatment experiment offers only one treatment comparison, whereas a three-treatment experiment offers three comparisons (Fig. 5-2). Hence, adding a third treatment to a two-treatment experiment can triple its value from a treatment comparison point of view. Similarly, a four-treatment experiment has twice as many comparisons as a three-treatment experiment (six vs. three) and is 50% more efficient.

TREATMENTS	1	2	3	4	5
COMPARISONS	0	1	3	6	10

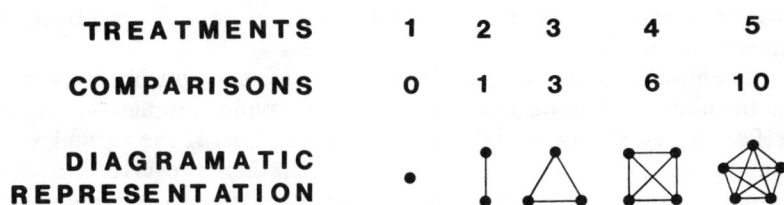

DIAGRAMATIC REPRESENTATION

Fig. 5-2. Diagramatic representation of the relationship between number of treatments and number of treatment comparisons (dots represent treatments and lines joining dots represent individual comparisons).

Practical Considerations

The practical value of grazing experiments refers to the extent results benefit the producer. This depends on the importance of the research topic relative to producer needs, and is clearly related to biological value. Furthermore, since livestock production is primarily a commercial activity, the practical value largely concerns the suitability of grazing research data for economic analysis. In this regard, if consistent with objectives, treatments should simulate production conditions as far as possible, and production functions need to be generated to facilitate use of economic optimization procedures. The key element of design here is the need for multiple (as opposed to single) levels of economically important continuous variables (variables that can be quantified, such as stocking rate, level of fertilizer, etc.) in each experiment (Jacobs, 1974).

Statistical Considerations

The statistical value of grazing experiments concerns the suitability of data for appropriate statistical analysis. It determines reliability of results and the level of confidence with which inferences about both the biological and practical aspects can be extended to the populations under study. As such, statistical considerations are essential in all grazing studies and should be given very careful attention at the planning stage. However, statistical considerations must always be strongly linked to the biology of the system under study.

The key statistical elements of design are those that ensure provision of a valid error term with adequate power for testing biologically and/or economically meaningful differences among treatments, and for determining confidence limits. These elements include randomization, the number of error degrees of freedom (df), number of paddocks, number of animals per paddock, and length of the grazing season. Implications of these factors relative to analytical procedures and control of experimental error have been mostly documented (e.g., Mott & Lucas, 1952; Petersen & Lucas, 1960; Owen & Ridgeman, 1968; Matches, 1970). Randomization of both paddocks and animals is, of course, a prerequisite for valid statistical analysis. Paddock size is determined by the need for each paddock to be stocked with preferably three to five animals throughout the grazing period. This will depend on many factors such as climate and expected productivity of treatments, heterogeneity among animals, and vulnerability of animals to adverse environmental conditions.

Unintentional confounding of important variables in grazing experiments should be minimized, but is nearly impossible to avoid completely. Scarnecchia (1988) provides a helpful discussion on this topic. If the variables confounded in a particular design are identified, then the effects of such confounding can be estimated. Although this is mostly a value judgement, it will often lead to clear acceptance or rejection of a potential design.

The general principle involved in all statistical tests is comparison of variation that can be accounted for (e.g., by treatments, blocks, regression coefficients, etc.) with variation that cannot be accounted for (the error term). If the former is sufficiently large relative to the latter, statistical significance is indicated. For designs that include replication, the error term is the variance among experimental units treated alike (replicates) or deviations from treatment means. This error term is preferred for tests of statistical significance because it provides the most reliable estimate of variance associated with a single mean. However, it appears that no error term is perfect. For example, Snedecor and Cochran (1967) point out that even the error term provided by replication assumes additivity (and no treatment × replicate interaction), and that this assumption is not assured. With the regression approach, which makes use of several nonreplicated grazing levels per treatment (Riewe, 1961; Bransby et al., 1988), regression lines can be considered as two-dimensional means (as opposed to a one-dimensional mean or point). In a sense, treatments are replicated, with replicates deliberately but randomly confounded with grazing level. Pooled deviations from regression is the appropriate error term (Draper & Smith, 1966; Zar, 1984). Both error terms (deviations from means or deviations from regression) have paddock and animal sources of variation confounded. Pooled deviations from regression also includes nonlinear effects, which are almost sure to exist. If these are strong, statistical tests will be more conservative. However, numerous studies in the literature suggest that the linear model is a good approximation, provided an appropriate range in grazing levels is chosen (e.g., Riewe, 1961; Cowlishaw, 1969; Hart, 1972; Jones & Sandland, 1974; Sandland & Jones, 1975; Hart, 1978; Bransby et al., 1988).

Because of general restriction on the number of experimental units, a shortage of error df and associated lack of power for statistical tests is often a major weakness in grazing studies. Consequently, the number of error df is one of the most important features of design, and its influence on test statistics (such as the F or t statistic) is of particular interest (Fig. 5-3). Increasing

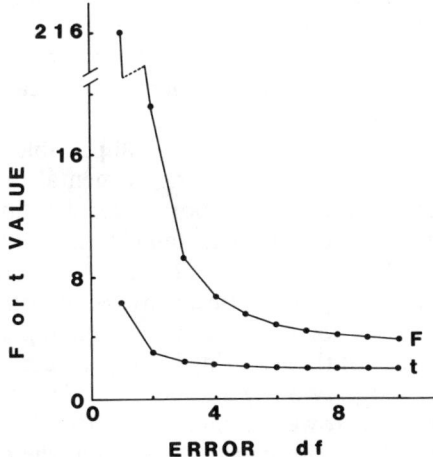

Fig. 5-3. Values of the F and t statistics for different error df, assuming three numerator df for the F ratio.

error df from 2 to 3, and from 5 to 6, results in reductions in the F statistic of 52 and 12%, respectively, and corresponding reductions in the t statistic of 19 and 4%. Implications of this for grazing experiments that have very low paddock numbers are critical, because of extreme sensitivity of test statistics to error df at the lower end of the error df range. For example, if a six-paddock grazing facility is to be used for an experiment with three treatments and two replicates, a randomized complete block design will result in only 2 error df, but a completely random design offers 3 error df. If a randomized block design is used and the block effect is small, considerable power is lost relative to a completely random design. Yet, it is not possible to predict block effects prior to the experiment. Perhaps the best approach is to use a completely random design with rectangular paddocks oriented in the direction of known or expected variation to minimize experimental error. Recognizing the severe restriction placed on power of statistical tests by limited error df, grazing experiments with less than eight paddocks and 4 error df seem hard to justify, regardless of design.

IDENTIFYING THE DESIRED COMPROMISE

The intended strategy in designing experiments is to test treatment differnces and/or establish biological principles. Financial and logistical constraints make cost-effectiveness important, forcing a compromise among the key elements of design (number of treatments, replicates, stocking levels, etc.) relative to objectives, and influencing the balance among biological, practical, and statistical aspects. Because of these constraints, modeling and computer experiments could make a particularly valuable contribution to the understanding of grazing ecosystems if properly integrated with field research. Therefore, the specific needs of modeling specialists relative to experimental design should be considered whenever possible. Selecting an appropriate compromise requires identifying all possible designs for a fixed set of research facilities, establishing priorities, and implementing a systematic elimination procedure. It is important to strive for sound biological and statistical logic throughout this process.

The first steps are to identify and rank all possible experimental objectives and to determine upper limits on experimental facilities, particularly paddock and animal numbers, and labor. After determining the paddocks available (12 in this example), the next step is to list all feasible designs and associated details of their key elements (Table 5-1). This makes it easy to narrow the choice to a few possibilities, provided constant attention is paid to prioritized objectives. For example, if maximizing treatments is a high priority, and a minimum of three grazing levels per treatment is desired (based on arguments presented previously and not considering partially replicated designs), the choice is narrowed to Options 6, 10, 11, and 12 (Table 5-1). Option 10, like Option 7, is unrealistic because of the excessive number of grazing levels. Generally, three grazing levels will be adequate if replicated, but without replication four would be preferable. All other options have either

DESIGN & CONDUCT OF GRAZING EXPERIMENTS

Table 5-1. Relationships among number of treatments (T), grazing levels (GL), replicates (reps.), treatment comparisons (TC), and degrees of freedom (df) allocations for a range in feasible design options suitable for a 12-paddock grazing research facility.

Design option	T	GL	Reps.	TC	Analysis Replicated T	GL	Reps.	Error	Nonreplicated regression Slopes	Inter- cepts	Error
								df			
1	6	1	2	15	5	0	1	5	--	--	--
2	4	1	3	6	3	0	2	6	--	--	--
3	3	1	4	3	2	0	3	6	--	--	--
4	2	2	3	1	1	1	2	7	--	--	--
5	3	2	2	3	2	1	1	7	--	--	--
6	2	3	2	1	1	2	1	7	--	--	--
7	1	6	2	0	0	5	1	5	--	--	--
8	1	4	3	0	0	3	2	6	--	--	--
9	1	3	4	0	0	2	3	6	--	--	--
10	2	6	1	1	--	--	--	--	2	2	8
11	3	4	1	3	--	--	--	--	3	3	6
12	4	3	1	6	--	--	--	--	4	4	4

too few treatments or too few grazing levels. Replicates severely deplete treatments and/or grazing levels, resulting in three out of four remaining options being nonreplicated designs.

The next choice is between replicated (Option 6) and nonreplicated (Options 10, 11, and 12) designs. In multiple grazing level experiments, replication can serve several functions. Of particular importance is the testing of regression relationships for linearity, and fitting nonlinear regressions if necessary. This need is most likely to occur with species mixtures, such as with grass-legume pastures (Jones, 1981), but is less likely on rangeland (Hart, 1978) and on single-species pastures, provided grazing levels are appropriate. Replication will also allow differences between individual grazing levels to be tested. This cannot be done with the regression approach, although significant slopes of regression lines indicate significant grazing level effects. Consequently, Option 6 would be the design of choice if a nonlinear response to grazing level was expected, or if testing differences between grazing levels was a high priority. However, replication occurs at the expense of one or two treatments and two or five treatment comparisons relative to Options 11 and 12, respectively.

When a linear response to stocking level is expected, nonreplicated designs and regression analysis can be used, particularly if the experimental site is uniform. Under such conditions the cost of replication (in terms of lost biological information associated with fewer treatment comparisons) is extremely high and difficult to justify. The biggest difference between Options 11 and 12 is in the number of treatment comparisons. If four treatments are of equal or near-equal importance, such as with a species or variety comparison, then Option 12 might be selected. Alternatively, if only three treatments are of primary interest, Option 11 can be used. Despite low error df for Option 12, it is unlikely that slopes of all four regression lines will be significantly different from one another, thus allowing error df to be increased (Bransby et al., 1988). If year-to-year carry-over effects are not of interest, it is possible to rerandomize each year and replicate in time (Burns et al., 1970). However, without rerandomization, this procedure is not valid, and if large year differences occur among treatments, the error term will be inflated.

Another general option would be to use only one grazing level per treatment, four or six treatments, and three or two replicates (Options 2 and 1 in Table 5-1), respectively. Without multiple grazing levels the treatment response becomes a single point on the grazing level continuum, instead of a response function. The choice between this alternative and Options 11 and 12 amounts largely to preference of the researcher to have a single point relatively well-defined, or a response function perhaps less well-defined. This will depend on the assessed importance of biological, practical, and statistical aspects of the experiment relative to objectives.

Compared to Option 11, Option 5 may appear attractive. Both options have three treatments, but Option 11 sacrifices two replicates (and the associated estimate of true error) for two (probably intermediate) grazing levels. Conversely, Option 5 sacrifices the two intermediate grazing levels so that the two extreme levels can be replicated twice. This decision becomes one of biological vs. statistical considerations. It is reasonable to assume that most true responses to grazing level are not linear, but that a linear model provides a good approximation of the range in grazing level that is of economic importance. In some cases this range may be narrow, whereas in others it may be quite wide, but this information is seldom available prior to an experiment. Therefore, in the case of two extreme grazing levels per treatment one runs the risk of grossly misrepresenting a truly nonlinear response, whether replication is present or not (Fig. 5-4). This could be just as misleading as use of only one grazing level in the situation described in Fig. 5-3. Consequently, the biological and practical (producer related) risk of only two replicated but extreme grazing levels per treatment appears to be high. Statistical risk associated with Option 11 is likely to be low given that (i) measurements of herbage mass levels prior to grazing an experiment provide an estimate of variation in productivity among paddocks at that time; (ii) experiments are repeated for at least 2 to 3 yr, with different animals each year; (iii) between-paddock variation is usually low relative to between-animal variation; and (iv) important sward measurements such as herbage mass and species composition should be taken regularly during the grazing period. Furthermore, Option 5 provides no paddocks in the grazing level region that is most likely to be optimum. These arguments tend to favor Option 11.

Modern computer capability would probably be best exploited by proper integration of simulation modeling and computer experiments with field research. Effective modeling requires expression of treatment responses as regression lines or response surfaces, rather than single means or points. Unfortunately, if data are intended for modeling and economic analysis, this will probably have to be at the expense of formal replication. However, regres-

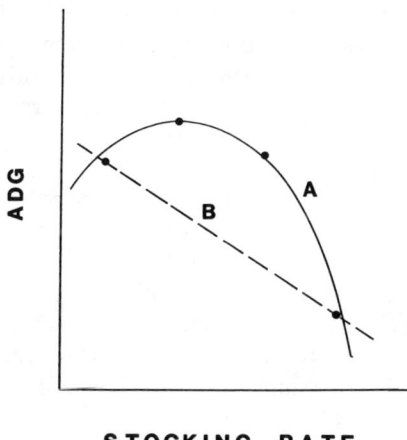

Fig. 5-4. Misrepresentation of a hypothetical nonlinear response (A) of average daily gain (ADG) to stocking rate by a linear function (B) based on the two extreme stocking rates.

sion procedures allow valid statistical analysis of multiple grazing-level experiments without formal replication. Therefore, under appropriate conditions this design appears to be justified if consistent with objectives, and is very competitive in cost-effectiveness.

Finally, designing and conducting a grazing experiment is similar to designing and building a house. If the foundations are weak, the whole structure will be weak (no matter how well the walls and the roof are built) and will fall apart under stress. A grazing experiment must be designed and conducted to stand up to the stress of experimental error while adequately pursuing stated objectives. The ingredients of the foundation are therefore those elements of design that strongly influence validity and the magnitude of the error term. Information in the literature suggests that the following factors and associated minimum standards will provide a solid foundation for grazing experiments: (i) randomization of animals and paddocks, (ii) four error df, (iii) eight paddocks, (iv) two to five animals per paddock, (v) a grazing period of 60 to 90 d, and (vi) at least two environments (years or locations). Because of interactions among these factors, the assigned values can be regarded only as guidelines. However, if these elements of design are collectively deficient, the whole experiment will be deficient and cannot be rescued by other experimental components like replication or method of stocking (P&T vs. fixed). Provided commonly accepted foundation ingredients of a grazing experiment (items i–vi above) are adequate, approaches that accommodate different objectives (e.g., use of replication vs. regression, or fixed vs. variable stocking) should be optional.

CONCLUSIONS

Financial and logistical constraints are often blamed for less than optimum mix among key elements of experimental design, and for inefficient use of facilities available for grazing research. Because little can usually be done about facility constraints, opportunity to increase cost-effectiveness of grazing studies relative to stated objectives lies mainly in alternative experimental designs and computer modeling. Comparatively little attention has been devoted to alternative designs (e.g., Riewe, 1961; Matches et al., 1974; Burns et al., 1983) in the last 30 yr. Mochrie (1981) indicated that 77% of the grazing studies included in a U.S. survey used randomized complete block or completely random designs. It is possible that in many of these cases (depending on objectives) alternative designs would have been more efficient than traditional designs, and equally appropriate. However, grazing researchers might not be qualified to make this judgement alone or to develop alternative designs; partially replicated designs also need to be evaluated. Therefore, productive partnerships with statisticians should be helpful in pursuing grazing research goals.

REFERENCES

Bransby, D.I. 1985. Modelling grazing intensity studies. p. 1092–1093. *In* Proc. 15th Int. Grassl. Congr., Kyoto, Japan. 24–31 Aug. 1985. The Natl. Grassl. Res. Inst., Nishi-narino, Japan.

Bransby, D.I. 1988. Rotational and continuous grazing interactions with stocking rate on warm season perennial pastures. p. 97–101. *In* Proc. For. and Grassl. Conf., Baton Rouge, LA. 11–14 April. AFGC, Belleville, PA.

Bransby, D.I., B.E. Conrad, H.M. Dicks, and J.W. Drane. 1988. Justification for grazing intensity experiments: analyzing and interpreting grazing data. J. Range Manage. 41:274–279.

Burns, J.C., F.G. Giesbrecht, R.W. Harvey, and A.C. Linnerud. 1983. Central appalachian hill land pasture evaluation using cows and calves. I. Analysis for an unbalanced grazing experiment. Agron. J. 75:865–871.

Burns, J.C., R.D. Mochrie, H.D. Gross, H.L. Lucas, and R.Teichman. 1970. Comparison of set-stocked and put-and-take systems with growing heifers grazing coastal bermudagrass (*Cynodon dactylon* L. Pers.). p. 904–909. *In* M.J.T. Norman (ed.) Proc. 11th Int. Grassl. Congr., Queensland, Australia. 12–23 April. Univ. of Queensland Press, St. Lucia, Queensland, Australia.

Burns, J.C., R.D. Mochrie, and D.H. Timothy. 1984. Steer performance from two perennial *Pennisetum* species, switchgrass, and a fescue–Coastal bermudagrass system. Agron. J. 76:795–800.

Cowlishaw, S.J. 1969. The carrying capacity of pastures. J. Br. Grassl. Soc. 24:207–214.

Draper, M.R. and H. Smith. 1966. Applied regression analysis. John Wiley & Sons, New York.

Hart, R.H. 1972. Forage yield, stocking rate, and beef gains on pasture. Herb. Abstr. 42:345–353.

Hart, R.H. 1978. Stocking rate theory and its application to grazing on rangelands. p. 547–550. *In* D.N. Hyder (ed.) Proc. 1st Int. Rangeland Congr., Denver, CO. 14–18 August. Society for Range Management, Denver, CO.

Jacobs, V.E. 1974. Needed: A systems outlook in forage-animal research. p. 33–48. *In* R.W. van Keuren (ed.) Systems analysis in forage crops production and utilization. CSSA Spec. Publ. 6. CSSA, Madison, WI.

Jones, R.J. 1981. Interpreting fixed stocking rate experiments. p. 419–430. *In* J.L. Wheeler and R.D. Mochrie (ed.) Forage evaluation: Concepts and techniques. AFGC/CSIRO, Melbourne, Australia.

Jones, R.J., and R.L. Sandland. 1974. The relation between animal gain and stocking rate. J. Agric. Sci. 83:606–611.

Marten, G.C., and R.M. Jordan. 1972. Put-and-take vs. fixed stocking for defining three grazing levels by lambs on alfalfa-orchardgrass pastures. Agron. J. 64:69–72.

Matches, A.G. 1970. Pasture research methods. p. I-1–I-32. *In* R.F Barnes et al. (ed.) Proc. Natl. Conf. Forage Qual. Eval. Utiliz., Lincoln, NE. 3–4 Sept. 1969. Nebraska Center for Continuing Education, Lincoln, NE.

Matches, A.G., F.A. Martz, D.A. Sleper, and M.T. Krysowaty. 1981. Selecting levels of herbage allowance to compare forages for animal performance. p. 331-340. *In* J.L. Wheeler and R.D. Mochrie (ed.) Forage evaluation: Concepts and techniques. AFGC/CSIRO, Melbourne, Australia.

Matches, A.G., F.A. Martz, and G.B. Thompson. 1974. Multiple assignment tester animals for pasture systems. Agron. J. 65:719-722.

Matches, A.G., and G.O. Mott. 1975. Estimation of the parameters associated with grazing systems. p. 203-208. *In* R.L. Reid (ed.) Proc. Third World Conf. Anim. Prod., Melbourne, Australia. 22-30 May 1973. Sydney Univ. Press, Sydney, Australia.

McLaren, J.B., R.J. Carlisle, H.A. Fribourg, and J.M. Bryan. 1983. Bermudagrass, tall fescue and orchardgrass combinations with clover or N fertilization for grazing steers. I. Forage growth and consumption, and animal performance. Agron. J. 75:587-592.

Mochrie, R.D. 1981. Survey of techniques used in grazing trials in U.S. from 1975 to 1980. p. 449-459. *In* J.L. Wheeler and R.D. Mochrie (ed.) Forage evaluation: Concepts and techniques. AFGC/CSIRO, Melbourne, Australia.

Mott, G.O. 1960. Grazing pressure and the measurement of pasture production. p. 606-611. *In* Proc. 8th Int. Grassl. Congr., Reading, England. 11-21 July. Alden Press, Oxford, England.

Mott, G.O., and H.L. Lucas. 1952. The design, conduct, and interpretation of grazing trials on cultivated and improved pastures. p. 1380-1385. *In* Proc. 6th Int. Grassl. Congr., State College, PA. 17-23 August. Pennsylvania State Univ., State College, PA.

Owen, J.B., and W.J. Ridgeman. 1968. The design and interpretation of experiments to study animal production from grazed pasture. J. Agric. Sci. 71:327-335.

Petersen, R.G., and H.L. Lucas. 1960. Experimental errors in grazing trials. p. 747-750. *In* Proc. 8th Int. Grassl. Congr., Reading, England. 11-21 July. Alden Press, Oxford, England.

Petersen, R.G., and H.L. Lucas. 1968. Computing methods for evaluating pastures by means of animal response. Agron. J. 60:682-687.

Read, J.C., and B.J. Camp. 1986. The effect of the fungal endophyte *Acremonium coenophialum* in tall fescue on animal performance, toxicity and stand maintenance. Agron. J. 78:848-850.

Riewe, M.E. 1961. Use of the relationship of stocking rate to gain of cattle in an experimental design for grazing trials. Agron. J. 53:309-313.

Riewe, M.E. 1984. Use of fixed and variable grazing management in pasture evaluation. p. 61-84. *In* C. Lascano and E. Pizarro (ed.) Evaluation of pastures with animals: Alternative methodologies. Int. Center of Tropical Agric. (CIAT), Cali, Colombia.

Sandland, R.L., and R.J. Jones. 1975. The relation between animal gain and stocking rate in grazing trials: An examination of published theoretical models. J. Agric. Sci. 85:123-128.

Scarnecchia, D.L. 1988. Minimizing confounding in case studies of agricultural systems. Agric. Syst. 26:89-97.

Snedecor, G.W., and W.G. Cochran. 1967. Statistical methods. Iowa Stae Univ. Press, Ames, IA.

Sollenberger, L.E., W.R. Ocumpaugh, V.P.B. Euclides, J.E. Moore, K.H. Quesenberry, and C.S. Jones, Jr. 1988. Animal performance on continuously stocked 'Pensacola' bahiagrass and 'Floralta' limpograss pastures. J. Prod. Agric. 1:216-220.

Wheeler, J.L. 1962. Experimentation in grazing management. Herb. Abstr. 32:1-7.

Wheeler, J.L., J.C. Burns, R.D. Mochrie, and H.D. Gross. 1973. The choice of fixed or variable stocking rates in grazing experiments. Exp. Agric. 9:289-302.

Zar, J.H. 1984. Biostatistical analysis. Prentice-Hall, Englewood Cliffs, NJ.

6 Compromises and Statistical Designs for Grazing Experiments

J. Wanzer Drane
University of South Carolina
Columbia, South Carolina

ABSTRACT

Research designs suitable for grazing studies are many and varied. The ability to detect treatment differences depends entirely on the model of the grazing experiment and the research design actually used. Small changes in the design can increase or decrease the sensitivity of a test by virtue of the mean square used for the error variance. Examples of experiments are evaluated and compared herein from the point of view of statistical designs and analysis. Invariably, the final design is a compromise between an ideal and the realities of budget and time constraints.

In row-crop and grain experiments, land is divided into plots. Treatment variables are applied to the plots, and at the appropriate times yields are measured. If yield is measured as a single number for a plot, then random variability among plots cannot be measured directly without replicating the experiment. Let us consider a randomized block experiment wherein strips of land called *blocks* are subdivided into plots of nearly equal sizes. In fact they are almost always considered to be exactly the same size. Then treatment variables are randomly assigned to the plots within the blocks.

The statistical effects of the experiment under discourse are assignable to treatments and blocks and their interactions, and in most cases, both are considered fixed. The linear additive model, Eq. [1], and the analysis of variance (ANOVA) (Table 6-1) have the following forms:

$$Y_{ij} = M + B_i + T_j + BT_{ij} + E_{ij} \qquad [1]$$

wherein Y_{ij} is the response to the jth treatment applied to the ith block; M is the overall population mean; B_i is the block effect, and T_j is the treatment effect. If the treatment effects are not constant from block to block, then the term BT_{ij} is added. Random variation, E_{ij}, is omnipresent whether or not it can be measured.

Copyright © 1989 Crop Science Society of America and American Society of Agronomy, 677 S. Segoe Rd., Madison, WI 53711, USA. *Grazing Research: Design, Methodology, and Analysis*, CSSA Special Publication no. 16.

Table 6-1. Skeletal ANOVA for a two-way completely crossed experiment (randomized complete block).†

ANOVA

Source	df	SS contrast	Expected mean square	
Blocks	$I - 1$	$\overline{Y}_i. - \overline{Y}..$	$JS_i B_i^2/(I - 1)$	$+ V_e$
Treatment	$J - 1$	$\overline{Y}._j - \overline{Y}..$	$IS_j T_j^2/(J - 1)$	$+ V_e$
$B \times T$	$(I - 1)(J - 1)$	$Y_{ij} - \overline{Y}_i. - \overline{Y}._j + \overline{Y}..$	$S_i S_j BT_{ij}^2/[(I - 1)(J - 1)]$	$+ V_e$
Plots	$IJ - 1$	$Y_{ij} - \overline{Y}..$	–	
Error	0	NA‡	NA	
Total	$IJ - 1$	$Y_{ij} - \overline{Y}..$	–	

† S_i, $S_i S_j$, etc. indicates a sum over all values of i, i, j, etc., respectively. V_e is the error variance.
‡ NA = not available.

Blocks are usually ignored. To test the hypothesis that the treatment effect is nil or constant across all treatments, the variance ratio and mean square treatment (MST) to mean square error (MSE),

$$F = \text{MST/MSE} \qquad [2]$$

should be used and compared to the upper percentage points of an F with df (degrees of freedom) equal to $J - 1$ and $(I - 1)(J - 1)$. This is not possible because there is only one observation per plot, and the random variation between plots cannot be estimated. What is usually done (Steel & Torrie, 1980, p. 195; or Montgomery, 1984, p. 211) is replace mean square error with mean square block by treatment (MSBT) to get

$$F = \text{MST/MSBT} \qquad [3]$$

by assuming, correctly or not, that the $B \times T$ interaction is zero. Thus, Eq. [3] is a central F statistic only if T_j and BT_{ij} are zero for all combinations of i and j. The hypothesis of no treatment effect is rejected if the calculated variance ratio equals or exceeds the tabulated value for a given error rate, i.e., 0.05 or smaller.

If for any reason one knows or believes that the forgoing interactions BT_{ij} are not zero, then the experiment should be designed to include estimates of the error variance, V_e. Equation [1] would be rewritten to reflect the same, namely

$$Y_{ijk} = M + B_i + T_j + BT_{ij} + E_{k(ij)}, \qquad [4]$$

where k is added to index the replication of the jth treatment applied to the ith block; and the ANOVA table would also change as in Table 6-2.

Table 6-2. Skeletal ANOVA for a two-way completely crossed experiment with replications within every ij combination.

ANOVA

Source	df	SS contrast	Expected mean square	
Blocks	$I - 1$	$\overline{Y}_{i..} - \overline{Y}_{...}$	$JKS_iB_i^2/(I - 1)$	$+ V_e$
Treatment	$J - 1$	$\overline{Y}_{.j.} - \overline{Y}_{...}$	$IKS_jT_j^2/(J - 1)$	$+ V_e$
$B \times T$	$(I - 1)(J - 1)$	$\overline{Y}_{ij.} - \overline{Y}_{i..} - \overline{Y}_{.j.} + \overline{Y}_{...}$	$KS_iS_jBT_{ij}^2/[(I - 1)(J - 1)]$	$+ V_e$
Plots	$IJ - 1$	$\overline{Y}_{ij.} - \overline{Y}_{...}$	-	
Error	$IJ(K - 1)$	$Y_{ijk} - \overline{Y}_{ij.}$		$+ V_e$
Total	$IJK - 1$	$Y_{ijk} - \overline{Y}_{...}$	-	

If the denominator in Eq. [3] is replaced with the MSE, we get the following:

$$F \text{ (treatment)} = \text{MST/MSE, and } F(B \times T) = \text{MSBT/MSE} \quad [5]$$

One answer to the presence of interactions is replication within every treatment thereby allowing for an estimate of the error variance. Although there may be variability from plot to plot, in general there is no great concern expressed over it. This is not true with researchers in grazing experiments.

When blocking is not available, desirable, or less than artificial, the experiment may resolve to one completely randomized over experimental units. Let us consider one illustrated completely in Steel and Torrie (1980, p. 153).

Treatments are applied to the soil within pots within a greenhouse. Mint roots are planted in equal or different numbers in each pot. The resultant linear additive model for equal numbers of roots within all pots is

$$Y = M + T_i + P_{j(i)} + R_{k(ij)} \quad [6]$$

wherein, M is the grand mean; T_i is treatment effect, $P_{j(i)}$ is the between-pot variability, and $R_{k(ij)}$ is the between-root variability. The treatment effect (T_i) is externally imposed and considered—truly an effect fixed in nature. Both $P_{j(i)}$ and $R_{k(ij)}$ are random variables and measure variation from pot to pot and from mint (*Mentha arvensis* L.) plant to plant. The skeletal ANOVA is shown in Table 6-3.

Table 6-3. Skeletal ANOVA of the completely nested mint plant experiment.

ANOVA

Source	df	SS contrast	Expected mean square
Treatment	$I - 1$	$\overline{Y}_{i..} - \overline{Y}_{...}$	$JKS_iT_i^2/(I - 1) + KV_p + V_r$
Pot	$I - 1$	$\overline{Y}_{ij.} - \overline{Y}_{i..}$	$KV_p + V_r$
Root	$IJ(K - 1)$	$Y_{ijk} - \overline{Y}_{ij.}$	V_r
Total	$IJK - 1$	$Y_{ijk} - \overline{Y}_{...}$	

Because the pot mean square is the error term for testing one treatment against another, the ability to detect differences among treatments rests in the number of pots that can be properly maintained within the greenhouse, unless the number of plants within a pot is part of the treatment. Then, the number of roots per pot is dictated by the treatment itself and can be used as a concomitant variable to measure root pressure or root intensity. If, in addition to root pressure, the experiment were carried out in the open (not in a greenhouse), then a second concomitant variable could also be used, namely rainfall. These would alter Eq. [6] as follows:

$$Y_{ijk} = A + B_1 N_{ij} + B_2 W_{ij} + T_i + P_{j(i)} + E_{ijk} \qquad [7]$$

where A replaces M and is considered the intercept; B_1 is the regression coefficient for the number of roots per pot, N_{ij}; and B_2 is the regression coefficient for centimeters of rain, W_{ij}, received by the pots. The term $R_{k(ij)}$ is now replaced by E_{ijk}, the residual error after extracting SS(B_1, B_2) from SSR; SS(B_1, B_2) is the sum squares due to regressing Y_{ijk} on N_{ij} and W_{ij}, and SSR is the sum of squares due to (or between) pot variability. The sum of squared error, SSE, is used to test A, B_1, and B_2 and the sum of squares due to plants within pots, SSP, remains the error term to test T unless the following is true: suppose the soil within all pots comes from the same batch and the only differences, practically speaking, come from N_{ij} and W_{ij}, which are then surrogates for $P_{j(i)}$. Equation [7] is now

$$Y_{ijk} = A + B_1 N_{ij} + B_2 W_{ij} + T_i + E_{ijk} \qquad [8]$$

At this juncture, MSE is the error term for all comparisons; SSE = SSR − SS(B_1, B_2) + SSP, the sum of squared deviations from regression.

DESIGNS FOR GRAZING EXPERIMENTS

Each of the foregoing models will be recast into designs that could be used in grazing experiments. Strengths and weaknesses of each will be discussed from the viewpoint of efficiency of use of experimental materials. Lastly, the statistical power of the test will also be discussed. Except where noted otherwise, gains are per–unit area.

Design 1. Paddocks are IJ in number. They are assigned to I groups, which are as nearly homogeneous as possible and are called *blocks*. These paddocks do not have to be contiguously arranged. To each of the I blocks, J treatments are randomly assigned. A treatment, let us remember, is a combination of externally imposed conditions or applications. A treatment could consist of a particular combination of components taken from (i) fertilizer kind and rate applied to the soil, (ii) specie or variety of forage, (iii) supplemental feed, (iv) source of water for the soil, (v) stocking rate, and (vi) other possible factors. Stocking rate will be treated separately from all others because the grazing animal is the biological material on which we make our measurements, and number of heads per hectare can be part of the treatment itself.

1. If among the J treatments, stocking rate is held constant, Eq. [1] is the linear additive model. The response variable is weight gained per paddock or hectare, and animal-to-animal variation contributes nothing to the experiment. Treatments are tested using MSBT, and the power of the test depends entirely on df $= (I - 1)(J - 1)$.

2. If stocking rate is part of the treatment combination but is not treated as an interval measure, there is no change in the model.

3. If stocking rate is considered an interval measure, Eq. [1] is replaced by

$$Y_{ij} = A + Cf(S_{ij}) + B_i + T_j + E_{ij} \qquad [9]$$

where A is the intercept; C is the regression coefficient for $f(S_{ij})$, the response function for the intensity of stocking; and E_{ij} is the residual error. Note that $f(S_{ij})$ is quadratic if the response is kilograms per hectare, and linear if the response is kilograms per animal (Riewe, 1961).

Design 2. Paddocks are IJK in number. The designs are those in Design 1 with K replicates of each block × treatment combination. The linear additive model is that of Eq. [4], and the response variable remains total gain per paddock or normalized to kilograms per hectare. Equation [5] would be used for testing if stocking rate were considered a nominal measure, but linear additive model Eq. [9] would be used if a response function were used to measure effects of varying stocking rates.

Design 3. Treatments are I in number, and each is applied to J paddocks containing N_{ij} animals and receiving W_{ij} centimeters of water. Let $g(W_{ij})$ be an appropriate response function of the amount of water received by the jth paddock. The linear additive models are those of Eq. [6], [7], and [8]. Therefore,

$$Y_{ijk} = A + B_1 f(S_{ij}) + B_2 g(W_{ij}) + T_i + P_{j(i)} + E_{ijk} \qquad [10]$$

or

$$Y_{ijk} = A + B_1 f(S_{ij}) + B_2 g(W_{ij}) + T_i + E_{ijk} \qquad [11]$$

depending on whether S_{ij} and W_{ij} are considered adequate surrogates for paddocks $P_{j(i)}$. This model applies to both kilograms per hectare and kilograms per animal by choosing the appropriate formulation of $f(S_{ij})$ as cited in Eq. [9].

Design 4, a Complex Example. This hypothetical grazing experiment will consist of JK treatments composed of K stocking rates, together with J combinations of other treatment components. The entire experiment is repeated over I years. Instead of water available to the soil, forage available to the grazing animal will be measured. Weight gain will be measured on each animal from which other measures of production can be calculated.

Forage availability will be taken as a surrogate for paddock effect and its response to the T_j by S_k combination. Thus, we are mapping the complex interaction of paddock × treatment × stocking rate into forage availability, which is a concomitant regression variable with a causal effect on weight gain for the animal. The linear additive model is

$$G_{ijkl} = B_0 + B_1 F_{ijk} + Y_i + T_j + YT_{ij}$$
$$+ S_k + YS_{ij} + TS_{jk} + YTS_{ijk} + A_{l(ijk)} \qquad [12]$$

where

G_{ijkl} = weight gain per animal or per hectare in this case;
B_0 = intercept;
B_1 = regression coefficient, assuming linear dependence between forage availability and weight gain;
F_{ijk} = forage availability;
Y_i = year-to-year effect;
T_j = treatment (other than S_k) effect;
S_k = stocking rate effect;
$A_{l(ijk)}$ = animal effect; and
YT, YS, TS, and YTS are the respective interactions.

Both T_j and S_k are considered fixed effects because they are predetermined by the persons conducting the experiment, whereas both Y_i and $A_{l(ijk)}$ are random variables. Table 6-4 is the ANOVA for Eq. [12].

The sum of squares due to forage availability, SSF, is a partition of the sum of squares due to the three-way interaction of year, treatment, and stocking rate, SSYTS, since F_{ijk} is completely confounded with YTS_{ijk}, except

Table 6-4. Skeletal ANOVA giving degrees of freedom and identifying appropriate error mean square for testing various components of the linear additive model.†

	ANOVA	
Source	df	Error mean square
F_{ijk}	1	A
Y_i	$I - 1$	A
T_j	$J - 1$	YT
YT_{ij}	$(I - 1)(J - 1)$	A
S_k	$K - 1$	YS
YS_{ik}	$(I - 1)(K - 1)$	A
TS_{jk}	$(J - 1)(K - 1)$	YTS
YTS_{ijk}	$(I - 1)(J - 1)(K - 1) - 1$	A
$A_{l(ijk)}$	$IJK(L - 1)$	--
Total	$IJKL - 1$	--

† L is the average number of animals over all paddocks.

that F_{ijk} is a continuous regression variable whereas YTS_{ijk} is discrete. Thus, the degree of freedom for F_{ijk} is subtracted from df for YTS_{ijk} and not $A_{l(ijk)}$.

Because the error mean squares for T_i and S_i are the YT and YS mean squares, respectively, this design has a major weakness. The role of the animal mean square is that of testing components of variance of Y and its interactions and the very important regression coefficient, B_1, of F_{ijk}. To gain respectable degrees of error to be used when testing T_j and S_k, it becomes necessary to rerun the exact same experiment, which usually takes several years.

An interesting collapse of Table 6-4 occurs for $I = 1$. That is, if one runs this experiment for only 1 yr, then Table 6-4 becomes Table 6-5.

The only random effect in this instant is animal, $A_{l(ijk)}$, and the power of the various tests depends on $JK(L - 1)$, the total number of animals JKL minus the number of paddocks JK. This can be a number of reasonable size and allow us to detect differences due to both stocking rates and the effects of other variables.

What is wrong with this design, and how can it be corrected? In Eq. [12] and Table 6-4 one can see the strong dependence on the number of years because the expected mean square (EMS) df is always that of the year × treatment interaction. This is the weakness. It can be corrected rather simply. Each year, retill all paddocks; reassign all JK combinations of T_j and S_k to the JK paddocks. Then each replicated experiment will be nested within year. This clearly has relevance only to annual pastures. Within Table 6-5 add a line for Y_i with df $I - 1$ and MSE = MSA, the mean square due to animal variation within year, treatment, and stocking rate. Multiply all other df by I, because the linear additive model is as follows:

$$Y_{ijkl} = B_0 + B_1 F_{ijkl} + Y_i + T_{j(i)} + S_{k(i)} + TS_{jk(i)} + A_{l(ijk)} \quad [13]$$

With this replicated experiment one cannot only obtain measures separating the T_i, S_k, and TS_{jk}, but year-to-year variability of the effects of forage availability F_{ijkl} can also be tested by testing for parallelism from year to year.

Table 6-5. Skeletal ANOVA giving degrees of freedom and identifying appropriate error mean squares for testing various components of the linear additive model.

	ANOVA	
Source	df	Error mean square
F_{jk}	1	A
T_j	$J - 1$	A
S_k	$K - 1$	A
TS_{jk}	$(J - 1)(K - 1) - 1$	A
$A_{l(jk)}$	$JK(K - 1)$	--
Total	$JKL - 1$	

CRITIQUE OF SOME PUBLISHED DESIGNS

Riewe (1961) reviewed a number of past trials giving their results and setting forth a design in which replication was not used or needed; his Table 1 gives 14 correlations between stocking rate and gain per animal. Most of the correlations have only 1 df for error. An $r = -0.999$ has a significance of only 0.028473 with df = 1. However, the sum of $-2\log(p)$ over the 14 tests of significance on the correlations is a chi-square with df = 28 (Sokal & Rohlf, 1981, p. 779). In this case it results in a chi-square equal to 96.95 and a level of significance of 1.7 in a billion. The evidence is very strong that average gain per animal can be expressed as a linear function of stocking rate.

When Riewe (1961) considers replication, whether on purpose or not, he treats a replication as a block and uses the block × treatment interaction as error. His Table 3 with F, P, and X^2 added and its replacements are Tables 6-6a and 6-6b, respectively.

The linear additive model used is that of Eq. [1] when it should have been that of Eq. [6] without the $R_{k(ij)}$ term. Here, he has lost 2 df for error, and the overall level of significance (p-value) was 4.5 times as large as it should have been. His claim that a linear relationship is adequate to express gain per animal (G) as a function of stocking rate is well-supported and leads to the following:

$$G = A + BS \text{ (kg/animal)} \qquad [14a]$$

$$= AS + BS^2 \text{ (kg/ha)} \qquad [14b]$$

and

$$S_{max} = -0.5A/B \qquad [14c]$$

where A and B are regression coefficients, S is stocking rate, and B is negative.

Table 6-6a. Riewe's Table with F/P, chi-square, and overall significance added.

Source	df	Mean square, 1957	F/P	Mean square, 1958	F/P
Reps	2	264	--	1969	--
Stocking rate	1	5891	38.5	7350	8.06
Error	2	153	0.025	912	0.105
	$X_4^2 = 11.89$			$Pr > X_4^2 = 0.018$	

Table 6-6b. Replacement of Riewe's Table 3 with F/P, chi-square, and overall significance added.

Source	df	Mean square, 1957	F/P	Mean square, 1958	F/P
Stocking rate	1	5891	28.3	7350	5.10
Reps	4	208.5	0.006	1440.5	0.087
	$X_4^2 = 15.12$			$Pr > X_4^2 = 0.004$	

Petersen et al. (1965) developed a theory linking stocking rate and per animal and per area performances, but do not address designs. Their initial assumption is not realistic to me: "Amount and type of forage available per acre are independent of stocking rate." In fact, as stocking rate increases, forage availability decreases (Conrad et al., 1981).

Conniffe (1976) sets out to compare between-herd and within-herd variances by collecting data from 12 experiments where both of these could be estimated. His model for error is

$$E_{ij} = A_{ij} + C_{ij} + H_i \qquad [15]$$

with expected mean squares of

$$\text{EMS (between)} = V_a + r V_h \qquad [16a]$$

$$\text{EMS (within)} = V_a + V_c \qquad [16b]$$

where
A_{ij} = animal effect,
C_{ij} = competition, and
H_i = herd effect.

He argues away C_{ij} by summing it to zero over the herd. This is incorrect! The term C_{ij} is inexorably confounded with and is a part of the animal variability itself and should not be part of the model at all. Removing V_c in Eq. [16b] gives the correct EMS. What is more, the linear additive model for the data he reports is that of Eq. [6] in which $H_{j(i)}$ replaces $P_{j(i)}$ and $A_{k(ij)}$ replaces $R_{k(ij)}$. In every case that he reports in his Table 3, the F test should be a one-tailed, right-tailed test. He incorrectly uses a two-tailed test in every case. His Table 3 (see Table 6-7) is here reproduced with the correct value of the F statistic and a column added for the probability of a larger F, where

$$F = \text{Between-herd MS} \div \text{Within-herd MS}$$

Table 6-7. Conniffe's Table 3 with abbreviated Experiment column, corrected Fs and $Pr > F$ added.

Experiment	Between herd	df	Within herd	df	F	Pr > F
1963, Moorepark	1770	2	704	76	2.33	0.10421
1964	4200	2	514	76	8.17	0.00061108
1965	664	2	421	78	1.58	0.21250
1966	58	2	638	78	0.091	0.91311
1967	271	2	734	79	0.369	0.69261
1968	1472	2	663	34	2.22	0.12411
1969	82	2	422	80	0.194	0.82404
1970	1275	3	687	131	1.86	0.13953
1967, Grange	145	3	345	29	0.420	0.74002
1968	150	3	323	30	0.464	0.70954
1969	89	3	337	30	0.264	0.85077
1970	171	4	533	36	0.321	0.86204

Again, the sum of $-2\log(Pr > F)$ is a chi-square with df equal to twice the number of independent tests of significance. In this case, $X^2 = 33.745$, df = 24, and $Pr > X^2 = 0.0893$. This alone is enough to fail to reject the hypothesis that the between-herd component of variance is zero. But Coniffee (1976) impeaches his 1964-Moorepark data, the one very large between-herd estimate of the variance. If we exclude that test, than $X^2 = 18.944$, df = 22, and $Pr > X^2 = 0.649$. This would lead me to conclude that the between-herd variance is either negligible or simply nonexistent.

COMPROMISES FOR GRAZING EXPERIMENTS

In this section, several statistical designs will be considered, which are available to the researcher who wishes to measure the effects of more than one treatment and more than one stocking rate with and without replication. Bransby et al. (1988) set forth convincing arguments for grazing each treatment in a grazing experiment at several intensities or stocking rates, regardless of whether replication were present. Everything has its price! From time to time compromises must be made between what is wanted and what is available.

The manner in which the experiment is designed and executed determines what constitutes the experimental unit, the proper error terms in the analysis of variance, and whether replication is either possible or desirable. In one of the earliest papers on the design of grazing trials, Mott and Lucas (1952) indicated treatments, precision or sensitivity, and cost as the three controlling factors; because of that, cost-based compromises may have to be made. Another early paper on design of grazing experiments is that of Matches (1969). Fundamentals of statistical methodology change slowly, even if our technologies change more rapidly. For example, we are now in a position to choose very sophisticated statistical designs, because computer technology allows us to analyze even the most complex data sets. Not very long ago computational constraints restricted such options.

If the objective of a grazing experiment is to evaluate two cultivars of two species, and each cultivar is fertilized at two levels of N and each is grazed at two stocking rates, then 16 paddocks will be required for a single replicate. The error mean square could come from higher-order interactions (three and four), because replication or partial replication would require more than 16 paddocks. The designs mentioned in grazing research literature include (i) randomized block, (ii) incomplete block, (iii) split plot, (iv) latin square, and (v) switch-back as well as the addition of covariables. Covariance analysis can accommodate continuous variables whether they are part of the design or not. One should plan carefully to utilize available resources to their fullest.

Matches (1969) reviewed much of the design work of Lucas, Mott, and Peterson and added his own experience in reducing experimental error by using halfsibs instead of animals pulled at random from one or more herds. Matches (1969) experienced gains in efficiency of 300 to 600% as measured by the coefficient of variation. What the reader may not realize when using the formulae of Mott and Lucas (1952) or Peterson and Lucas (1960) for calculating the experimental errors is that along with each formula is a specific statistical experimental design model. They are inseparable. (The grazing study is the experiment!) Details of the grazing experiment determine first the design and then the statistics available for analysis. The experimental error is calculated using the statistical model that reflects the experiment as it was carried out. By carefully considering all variables to be controlled within the study, the experimental error can be identified before the fact, and the investigator will have the opportunity to allocate resources against the most important criteria.

Assume 48 calves (*Bos* sp.) are assigned to 12 paddocks of equal size. What follows can easily be generalized to arbitrary numbers of animals and paddocks. Using specific numbers should enable the reader to follow more easily.

A Single-Factor Design. In a single-factor design, all variables are held constant except for the one of interest, and only that one variable will be reflected in the inferences available at the end of the grazing study. There can be no opportunity to infer an optimal combination of treatment variables for any purpose. If variety of a particular pasture specie is of interest, there can be no conclusion regarding the stocking rate that maximizes animal production per hectare or daily gain per animal on each variety. For such a design including 12 paddocks and 48 calves, the statistical model is

$$Y_{ijk} = M + V_i + P_{j(i)} + C_{k(ij)} \qquad [17]$$

where $i = 1, 2; j = 1, 6;$ and $k = 1, 4$ are the indices of variety, paddock within variety, and calf within paddock, respectively. Also, M is the grand mean, V is the variety effect, P is the paddock effect, and C is the calf effect. Both P and C are random variables and carry the burden of providing the error terms for testing hypotheses and calculating confidence intervals. Equation [17] is the statistical support for Table 6-8 and its implied analysis of variance. Most of the power of this design resides in the calf mean square with degrees of freedom equal to 36, and it is usually wasted on measuring the between-paddock variation instead of the difference between varieties. If the investigator does not reject the hypothesis that the between-paddock variability is equal to the between-calf variability, then the paddock and calf sums of squares can be combined (Hogg, 1961). Their combined mean squares

Table 6-8. Skeletal ANOVA for a single factor factorial with subsampling.

	ANOVA	
Source	df	Error mean square
Variety	1	Paddock
Paddock	10	Calf
Calf	36	--
Total	47	--

can be used to test for the variety effect with degrees of freedom for error being 46 instead of 10. This is contrary to Walker and Richardson (1986).

Suppose there are I treatments in a single factor grazing study, J paddocks per treatment, and K calves per paddock, and the paddock mean square is substantially larger than the calf mean square. We can still combine them in the following manner to increase the degrees of freedom for error when testing for treatment effect:

1. Add the paddock and calf sum of squares. The degrees of freedom will be $I(JK - 1)$ instead of $I(J - 1)$ as before.
2. Multiply this mean square by $(JK - 1)/(J - 1)$, but use the proper degree of freedom, namely $I(JK - 1)$, which you used as the divisor when calculating the mean square.

This combination produces an expected mean square of

$$\text{EMS}(P + C) = K\,V_p + [(JK - 1)/(J - 1)]\,V_c \qquad [18]$$

where V_p is the paddock variance and V_c is the calf variance. The EMS($P + C$) is slightly larger than the one using paddock mean square, but the degree of freedom for error is now $I(JK - 1)$ instead of $I(J - 1)$. In our example, df $= 2 \times (6 \times 4 - 1) = 46$ instead of 10. The increase in degrees of freedom should more than offset the slight inflation of the error term.

A Three-Factor Design. Suppose now on these same 12 paddocks we wish to test two varieties, three stocking rates, and two levels of N. The linear additive model for this experiment is

$$Y_{ijklm} = M + V_i + S_j + VS_{ij} + T_k + VT_{ik} + ST_{jk}$$
$$+ VST_{ijk} + P_{l(ijk)} + C_{m(ijkl)} \qquad [19]$$

and its ANOVA is indicated in Table 6-9. Y_{ijklm} is either kilograms per hectare or kilograms per animal.

Table 6-9. Skeletal ANOVA for a 2 × 3 × 2 factorial experiment on 12 paddocks holding a total of 48 calves.

	ANOVA	
Source	df	Error mean square
Variety	1	
Stocking rate	2	
VS	2	
Treatment	1	
VT	1	
ST	2	
VST	2	
Factor total	11	
Paddock	0	
Calves	36	
Total	47	

COMPROMISES & STATISTICAL DESIGNS

The randomization would be restricted so that there would be an average of four calves per paddock (e.g., 2, 4, and 6 per paddock). The levels of all the other variables would then be assigned at random to each set of paddocks at each stocking rate. Note that the degree of freedom for paddock is zero. This might lead one to conclude there is no legitimate error term for testing the experimental variables and their interactions. This is not so. There are several possible alternatives. For any factorial experiment with only one replicate, the treatments and plots are inexorably confounded. Usually investigators use the higher-order interaction sums of squares as error sum of squares, which is legitimate, if those interactions are in fact nonexistent. Here, we would have degrees of freedom equal to two, which would limit the power of the statistical test.

A second option is to test the three-way interaction with the calf sum of squares and, if found to be nonsignificant, combine the interaction and calf sum of squares (Hogg, 1961), and continue testing and combining until an interaction is found statistically significant. Continue testing, but do not combine any more interaction sum of squares with the then error sum of squares.

A third option is to sacrifice either one of the three variables and group the paddocks into blocks. Blocks are collections of nearly homogeneous plots, and they do not have to be contiguous to be members of the same block. Between-block variation should be large when compared to within block variation. All interaction terms involving blocks contribute to error sum of squares. Here, a replication of the entire set of treatments in the experiment is assigned to each block. There is not a within-treatment replication so often referred to in the literature in this design (Walker & Richardson, 1986, and others). Thus, replacing treatment with block and removing paddock from Table 6-2, we have Table 6-10.

The compromise is obvious. Giving up the variable, treatment bought an error term, but with degrees of freedom of only five. Five is small. What we may not see is that the interactions VB, SB, and VCB are pure error terms just as calf is, and the four terms should be combined giving a degree of freedom for error equal to 41. By blocking we lost the ability to make inferences about the treatments, but we gained statistical power.

Table 6-10. Skeletal ANOVA of a 2 × 3 factorial experiment on two blocks of six paddocks each.

ANOVA		
Source	df	Error mean square
Variety	1	$E1 + E2$
Stocking rate	2	$E1 + E2$
VS	1	$E1 + E2$
Block	1	--
$VB + SB + VSB = E1$	5	--
Factors + $E1$	11	--
Calf = $E2$	36	--
Total	47	--

Table 6-11. Skeletal ANOVA for a 2 × 3 + 2 factor factorial on 12 Paddocks plus forage availability.

ANOVA		
Source	df	Error mean square
Variety	1	Calf
Stocking rate	2	Calf
VS	2	Calf
Treatment	1	Calf
VT	1	Calf
ST	2	Calf
Forage avail	1	Calf
VST − F	1	Calf
Calf	36	--
Total	47	--

A Three-Factor Design with Covariate. To the above three-factor design ($2V \times 2S \times 2T$) let us add a measure of forage availability such as described by Bransby et al. (1988) and Hart (1972). Forage availability strongly affects animal gain, and if measured accurately, it can be taken as a surrogate for the paddock or pasture. The statistical model is that of Eq. [19] with the term BF added; B is the regression coefficient, and F is forage available in its proper units. Its ANOVA is that of Table 6-11.

In this experiment a three-factor factorial is applied in a completely randomized fashion to 12 paddocks. The only error term is the between-calf sum of squares, which has a reasonable degree of freedom and should be used to test all effects. The $VST - F$ (VST minus F) effect measures the interaction among variety, stocking rate, and treatment after the paddock effect has been removed via regression on forage availability, F. If this term is nonsignificant, all of the information contained in the data has been used.

DISCUSSION

The simplest experiment (single-factor factorial) is the most wasteful, information-wise, in that the paddock sum of squares constitutes the error term. To improve the power one has to resort to approximate methods that are quite acceptable, because the inflation of error mean square is due to the smaller (between-calf) component of variance. A single replicate multiway factorial design that uses all of the available paddocks completely removes the paddock variability as a disturbing influence on the research design, especially if the forage availability is also measured. The calf sum of squares emerges as the error sum of squares.

Other studies could be included for critiques, but that would be stretching the point. From my perspective, empirical forage research has advanced to the point that it is inseparable from that of mathematical and statistical modeling. There is an unlimited number of designs available to the grazing researcher. Whether replication of one kind or another must be used depends entirely on the modeling. Experimental materials are very expensive in this research, which goes almost without saying. A small quirk in the experiment as actually performed, against all good intentions, can result in loss of degrees of freedom in the error mean square and diminish the power of the test to the point of voiding the entire experiment.

Note that we can combine results of experiments through the use of $-2 \log(p)$, where p is the level of significance of a particular test of hypothesis. If, within K independent experiments, the same hypothesis is tested and an exact level of significance calculated, then the sum of $-2 \log(p)$ is a chi-square with degree of freedom $2K$. Pearson (1938), Littell and Folks (1973), Rosenthal (1978), and Berk and Cohen (1979) provide useful information for this and other means of combining tests. As a rule of thumb, if p is consistently less than $1/e$, then the combined level of significance can be made to have an arbitrarily small probability associated with it by increasing K. Combine enough weak results, and you can almost surely draw a strong conclusion. Caveat emptor!

REFERENCES

Berk, R.H., and A. Cohen. 1979. Asymptotically optimal methods of combining tests. J. Am. Stat. Assoc. 74:812-814.

Bransby, D.I., B.E. Conrad, H.M. Dicks, and J.W. Drane. 1988. Justification for grazing intensity experiments: Analysing and interpreting grazing data. J. Range Manage. 41:274-279.

Conniffe, D. 1976. A comparison of between herd and within herd variance in grazing experiments. Irrig. J. Agric. Res. 15:39-46.

Conrad, B.E., E.C. Holt, and W.C. Ellis. 1981. Steer performance on coastal, callie and other hybrid bermudagrasses. J. Anim. Sci. 53:1188-1192.

Hart, R.H. 1972. Forage yield, stocking rate, and beef gains on pasture. Herb. Abstr. 42:345-353.

Hogg, R.V. 1961. On the resolution of statistical hypotheses. J. Am. Stat. Assoc. 56:978-989.

Littell, R.C., and J.L. Folks. 1973. Asymptotic optimality of Fisher's method of combining independent test II. J. Am. Stat. Assoc. 68:193-194.

Matches, A.G. 1969. Pasture research methods. p. I-1-I-32. In R. F Barnes et al. (ed.) Proc. Natl. Conf. Forage Qual. Eval. Utilization. Lincoln, NE. 3-4 September. Nebraska Center for Continuing Education, Lincoln, NE.

Montgomery, D.C. 1984. Design and analysis of experiments. 2nd ed. John Wiley & Sons, New York.

Mott, G.O., and H.L. Lucas. 1952. The design, conduct, and interpretation of grazing trials on cultivated and improved pastures. p. 1380-1385. In Proc. 6th Int. Grassl. Congr., State College, PA. 17-23 August. Pennsylvania State Univ., State College, PA.

Pearson, E.S. 1938. The probability integral transformation for testing goodness of fit and combining independent tests of significance. Biometrika 30:134-148.

Petersen, R.G., and H.L. Lucas. 1960. Experimental errors in grazing trials. p. 747-750. In Proc. 8th Int. Grassl. Congr., Reading, England. 11-21 July. Alden Press, Oxford, England.

Petersen, R.G., H.L. Lucas, and G.O. Mott. 1965. Relationship between rate of stocking and per animal and per acre performance of pasture. Agron. J. 57:27-30.

Riewe, M.E. 1961. Use of the relationship of stocking rate to gain of cattle in an experimental design for grazing trials. Agron. J. 53:309-313.

Rosenthal, R. 1978. Combining results of independent studies. Psychol. Bull. 85:185-193.

Sokal, R.R., and F.J. Rohlf. 1981. Biometry. 2nd ed. W.H. Freeman and Co., San Francisco, CA.

Steel, R.G.D., and J.H. Torrie. 1980. Principles and procedure of statistics. 2nd ed. McGraw-Hill, New York.

Walker, J.W., and E.W. Richardson. 1986. Replication in grazing studies—why bother? p. 51-58. In C.D. Bonham (ed.) Symp. on Statistical Analysis and Modelling Grazing Systems Responses. 39th Annu. Meeting of the Society for Range Management, Orlando, FL. 10 Feb. Society for Range Management, Denver, CO.

7 Experimental Design and Statistical Inference: Generalized Least Squares and Repeated Measures over Time

F. G. Giesbrecht
North Carolina State University
Raleigh, North Carolina

ABSTRACT

Grazing experiments have invariably represented compromises between the ideal comprehensive experiment and the small trial resulting after concessions enforced by limited resources. Because factors having major effects generally are easier to discover, it follows that as progress continues there will be increasing emphasis on detecting ever smaller effects. Without a concomitant increase in size of experiments there will be an increased need to rely on conclusions gained from combining information from independent studies. This chapter explores some of the alternative designs and suggests methods for combining information from separate experiments.

An introductory comment by Petersen and Lucas (1960) stating that grazing experiments "are cumbersome and expensive" and that "they usually represent a compromise between the high costs of conducting an extensive trial on the one hand and the low sensitivity of an unnecessarily small trial on the other" is as true today as it was in the late 1950s. Increasing competition for scarce resources generally forces researchers into compromises or alternatives, which unfortunately might impinge on the ability to make sound inferences. One of the major changes that has come about since about 1975 has been the tremendous increase in our ability to perform complex arithmetical procedures with relative ease. Unfortunately this gain in flexibility has not aided experimental design as much as some would have expected. The fact still remains that if one wants to infer some sort of cause-and-effect relationship between experimental treatments and observed responses, then one must ensure that treatments are randomly assigned to the experimental units or in the case of grazing trials, treatments to paddocks. Also, one needs a measure of experimental error to evaluate whether the observed treatment

---

Copyright © 1989 Crop Science Society of America and American Society of Agronomy, 677 S. Segoe Rd., Madison, WI 53711, USA. *Grazing Research: Design, Methodology, and Analysis*, CSSA Special Publication no. 16.

differences are due to real effects or are due to chance alone, i.e., to distinguish real differences from inherent variation. The development of the computer, however, has lead to more sophisticated techniques for analysing data. These methods can be used to provide analyses of unbalanced or damaged experiments. Unfortunately, the resulting interpretations are never as direct or as definite as those obtained from balanced experiments that yield orthogonal contrasts. The value of the increased computing power and the associated new statistical techniques becomes apparent when one begins to analyze data sets obtained by merging results from diverse sources. Such merged data sets are almost invariably unbalanced and usually large. One object of this chapter is to examine some of the newer approaches to statistical data analysis and their relationship to the problems associated with grazing research.

One of the difficulties in grazing studies is that the experimenter is typically interested in measures related to quality, such as product per animal (average daily gain), and measures related to quantity, such as animal product per unit area. Although the two are related, they present some conflicts to the scientist planning his experiment. Mott and Lucas (1952) attribute the experimental errors (σ_{ee}^2) in grazing trials to two primary sources: (i) variations among pastures upon which the same experimental treatments have been imposed, and (ii) variations among animals subjected to the same environment. Petersen and Lucas (1960) then further subdivide the first error source into a component due to variations of the general quality level between pastures treated alike (σ_p^2) and a component due to variations of the time trend in quality level among pastures treated alike ($\sigma_{p \times t}^2$). They also partition the animal contribution to the experimental error per pasture in quality measures into two sources: (i) differences among animals as regards their general ability to produce (σ_a^2), and (ii) variations among animals as regards time trends ($\sigma_{a \times t}^2$). They also argue that in the context of product per animal experiments, the contribution to experimental error from the pasture component is independent of both the number of animals grazing and the length of time the pasture is grazed, the contribution from the pasture × time interaction should vary inversely as the length of time the pasture is grazed, the among animal contribution should vary inversely as the number of animals and the contribution from the animal × time interaction, inversely as the number of days grazed. All of these components can be combined into one expression:

$$\sigma_{ee}^2 = \sigma_p^2 + \frac{\sigma_{p \times t}^2}{t} + \frac{\sigma_a^2}{a} + \frac{\sigma_{a \times t}^2}{d} \qquad [1]$$

where t = length of grazing period, a = number of different animals grazing the pasture during the trial, and d = the number of animal-days the pasture is grazed.

In the context of animal per unit area experiments, there is a similar breakdown of the sources of error. However, they now argue that it is reasonable to assume that the plot contribution is related to the inverse of the square root of the plot size and the pasture × time interaction varies inversely as the number of days the pasture is grazed and inversely as the square root of the pasture size. The among animal contribution they claim should be expected to vary inversely as the number of animals and the contribution from the animal × time interaction inversely as the number of days grazed. The four components can again be combined into one expression:

$$\sigma_{ee}^2 = \frac{\sigma_p^2}{\sqrt{s}} + \frac{\sigma_{p \times t}^2}{t\sqrt{s}} + \frac{\sigma_a^2}{a} + \frac{\sigma_{a \times t}^2}{d} \qquad [2]$$

where s = pasture area, and as before t = length of grazing period, a = number of different animals grazing the pasture during the trial, and d = the number of animal-days the pasture is grazed.

Petersen and Lucas (1960) also report numerical estimates of the various components of variance, based on data from a rather extensive collection of grazing studies. One of their conclusions is that land area per pasture affects experimental error per pasture only for quantity measures, and even then its effect is relatively small. They recommend thinking in terms of animals carried rather than land area. They concluded, for example, that for pasture quality measures (product per animal studies), pastures with three to six animals appear to be nearly optimum size, whereas for quantity measures (animal product per unit area), pastures with one to three animals each appear to be nearly optimum. Pastures with two to five animals seem to be a good compromise. They also warn that the grazing period should be as long as is warranted by the condition of the sward. They feel that grazing periods as short as 2 to 5 wk are probably too short.

One can challenge several of the assumptions made by Petersen and Lucas (1960) in their derivation of Models [1] and [2]. Their paper makes no mention of attempts to examine Models that would result from different assumptions. A very worthwhile exercise would be to examine more recent data and test alternate hypotheses. For example, one could examine alternatives to the square root of pasture area term in the models, or see if land area has an impact on intake. An important point is that it is unlikely that a single investigator will ever have the resources to set up one experiment to test hypotheses of this sort. The only possible way to find answers will be to combine information from many studies.

POSSIBLE EXPERIMENTS

A brief scan of the literature seems to indicate that the typical grazing trial involves not more than 16 to 20 pastures and is repeated for 3 to 4 yr. A recent paper by Bransby et al. (1988) outlines some of the questions being asked about the design of grazing experiments, including the question of the need for replication. We now turn to some of these questions in terms of the context and method of approach established by the Petersen and Lucas (1960) and the Mott and Lucas (1952) work.

To be specific, consider a hypothetical experiment with 20 pastures, each suitable for three to four animals and available for 4 yr. The resulting data will consist of a 20 × 4 array of numbers (for a specific variable). On the basis of the objectives of the study and a critical review of the resources available, the experimenter must decide on the design to be used. Questions to be answered include the following:

1. Should the experiment be designed as a randomized complete block experiment, a simple completely randomized design, or possibly even a more complex design?
2. Should the treatments include a factorial structure involving several factors?
3. Should treatments be replicated (repeated in the conventional sense) in the experiment?

One of the simplest possible grazing experiments would be to assign t treatments, each to r pastures in a completely randomized design conducted for only 1 yr. This would lead to the analysis of variance illustrated in Table 7-1. One is immediately struck by the pasture × treatment interaction, denoted by $\sigma^2_{p \times t}$. It would be nice if one had assurance that this was negligible. Even if it does exist, however, it does not really destroy the validity of the experiment, but it does reduce the power of the experiment.

Then there is a question about what is in σ^2_e. Does the true σ^2_e contain only the sources listed for σ^2_{ee} in Eq. [1] and [2] or are there extra contributions due to relative performances of pastures and treatments varying over years? This experiment has sampled only 1 yr and consequently cannot by itself yield estimates of these components. It appears that there is a plausible argument that there really may be a true component of variation due to a year × treatment interaction, because some treatments may have a greater advantage (disadvantage) over other treatments in 1 yr than another ($\sigma^2_{yr \times t}$), and possibly also a true component of variation due to pasture × year interaction ($\sigma^2_{p \times yr}$). A pasture × year × treatment interaction, ($\sigma^2_{p \times yr \times t}$) could also be defined but for this discussion will be assumed negligible.

Rather than modifying the expected mean squares in the above analysis of variance table to reflect the extra components of variance, we modify the experiment by keeping the same t treatments on the r pastures for m years. This leads to the analysis of variance shown in Table 7-2. Note that although this analysis of variance reminds one of the split-plot experiment, it is an example of what many call a strip-block or split-block experiment with years forming strips in one direction and pastures the strips in the other direction. Obviously, treatments are randomly assigned to only the pasture strips and not the year strips. The component of variance due to pastures (σ^2_p) has been listed separately in this analysis because it does not occur in the treatment × year and the residual error line. These two are within pasture quantities. The Petersen and Lucas (1960) analysis indicated this component could be

Table 7-1. The expected mean squares for the analysis of variance of a completely randomized grazing trial with t treatments replicated r times in 1 yr.

Source	df	Expected mean square
Treatments	$t - 1$	$\sigma_e^2 + \sigma_{p \times t}^2 + \ldots$
Error	$t(r - 1)$	$\sigma_e^2 + \sigma_{p \times t}^2$
Total	$tr - 1$	

Table 7-2. The expected mean squares for the grazing trial with t treatments, each on r pastures, and continued for y yr.

Source	df	Expected mean square
Years	$m - 1$	$\sigma_e^2 + \sigma_{p \times yr}^2 + r\sigma_{yr \times t}^2 + rt\sigma_y^2$
Among pastures	$rt - 1$	
treatments	$t - 1$	$\sigma_e^2 + \sigma_{p \times yr}^2 + r\sigma_{yr \times t}^2 + m\sigma_{p \times t}^2 + m\sigma_p^2 + \ldots$
reps(trt)	$t(r - 1)$	$\sigma_e^2 + \sigma_{p \times yr}^2 + m\sigma_{p \times t}^2 + m\sigma_p^2$
Within pastures	$(rt - 1)(m - 1)$	
trt × years	$(t - 1)(m - 1)$	$\sigma_e^2 + \sigma_{p \times yr}^2 + r\sigma_{yr \times t}^2$
Error	$t(r - 1)(m - 1)$	$\sigma_e^2 + \sigma_{p \times yr}^2$
Total	$mtr - 1$	

ignored. Also, a possible pasture × year × treatment ($\sigma_{p \times yr \times t}^2$) interaction has been left out on the assumption that it is probably negligible.

Without further assumptions, this analysis provides a direct F test of the no year × treatment interaction hypothesis. A test of the no treatment effect hypothesis can be constructed using Satterthwaite's approximation (Steel & Torrie, 1980, p. 357). Alternatively, if there is no year × treatment interaction, then a valid F test exists.

Many researchers may feel that they cannot afford the luxury of repeating identical treatments on several pastures. In terms of the analysis presented, they prefer to set $r = 1$. This eliminates the rep(trt) and the error lines from the analysis of variance table presented. The experimenter, however, can still secure an estimate of experimental error provided the treatments are suitably selected. One possibility is to use a factorial structure, assume that some treatment interactions are negligible and then use the corresponding mean squares in the analysis of variance to estimate error. For example, in the hypothetical experiment considered above, the experimenter could set $r = 1$ and use a factorial treatment structure, say all combinations of h herbage mass levels, and c cultivars. The components of variance situation becomes complex. In particular, the year × treatment interaction ($\sigma_{yr \times t}^2$) and pasture × treatment interaction ($\sigma_{p \times t}^2$) components should be decomposed into components that involve herbage mass and cultivars. Keeping only the two factor interactions leads to an analysis of variance of the form shown in Table 7-3. Of course knowledge of the experimental material and the treatment factors may make it reasonable to drop more of the components of variance. Also a scientist may feel strongly that there should be a year × herbage mass × cultivar interaction component ($\sigma_{yr \times h \times c}^2$). This

Table 7-3. Expected mean squares for a factorial grazing experiment with h herbage mass levels and c cultivars repeated for m yr.

Source	df	Expected mean square
Years	$m - 1$	
Among pastures	$hc - 1$	
Herbage mass	$h - 1$	$\sigma_e^2 + \sigma_{p \times yr}^2 + m\sigma_p^2 + m\sigma_{p \times h}^2 + m\sigma_{p \times c}^2 + m\sigma_{h \times c}^2$ $+ c\sigma_{yr \times h}^2 + \ldots$
Among cultivars	$c - 1$	$\sigma_e^2 + \sigma_{p \times yr}^2 + m\sigma_p^2 + m\sigma_{p \times h}^2 + m\sigma_{p \times c}^2 + m\sigma_{h \times c}^2$ $+ h\sigma_{yr \times c}^2 + \ldots$
Interaction	$(h - 1)(c - 1)$	$\sigma_e^2 + \sigma_{p \times yr}^2 + m\sigma_p^2 + m\sigma_{p \times h}^2 + m\sigma_{p \times c}^2 + m\sigma_{h \times c}^2$
Within pastures	$(hc - 1)(m - 1)$	
Herbage mass $\times$ years	$(h - 1)(m - 1)$	$\sigma_e^2 + \sigma_{p \times yr}^2 + \sigma_{yr \times h}^2$
Cultivars $\times$ years	$(c - 1)(m - 1)$	$\sigma_e^2 + \sigma_{p \times yr}^2 + \sigma_{yr \times c}^2$
Error	$hc(m - 1)$	$\sigma_e^2 + \sigma_{p \times yr}^2$
Total	$hcm - 1$	

component will appear in each line of the expected mean squares with unit coefficient. The appropriate analysis of variance and corresponding expected mean squares for factorial grazing experiments involving three or more factors can be obtained by following the established pattern. Note that higher order factorial experiments are very attractive in the sense that they allow one to examine many factors in a modest number of pastures. Small single factor experiments may give misleading results because of the presence of unidentified interactions.

Blocking is another modification that is frequently introduced into grazing experiments. The guiding principle is to use blocking if there is a basis to group pastures (animals) into more homogeneous subsets. For example, in mountainous terrain it may be desirable to group pastures according to aspect exposure and then assign treatments so that all treatments occur on each exposure an equal number of times. Alternatively, breed or type of animal can be used as a blocking factor if the object of the experiment is to study animal product per unit area (pasture production).

ALTERNATIVE SOURCES FOR ERROR MEASURES

At this point two alternative methods for obtaining estimates of experimental error in grazing experiments have been examined. The first and probably the simplest is to repeat treatments. The second is to utilize the higher order interactions in factorial experiments. An old but very relevent discussion of the merits of factorial experiments in agricultural experiments can be found in Chapter 6 of Fisher (1953). Not only do factorial experiments allow the experimenter to study several factors with increased precision, the hidden replication effect, but they also provide a wider basis for inductive inferences. The measure of experimental error is obtained by assuming that higher order interactions are negligible and pooling the corresponding sums of squares in the analysis of variance. It unfortunately is not uncommon for researchers to shy away from factorial experiments because of the fear of having to deal with interactions. This is unfortunate, because it is not the experiment that gives rise to interactions. Interactions are a product of nature. If interactions are important and one conducts research one factor at a time, one will invariably be led to unwarranted conclusions. A factorial experiment at least permits a possible warning of underlying complexities. An additional point that is often missed is that although it is true that repetitions of a treatment (or treatment combination in the case of a factorial) provide a measure of experimental error, it does not follow that these repetitions always provide the proper error term to use for testing the treatments in an experiment.

There is a third alternative that is occasionally used to provide a measure of experimental error. The device is to plan the experiment in a manner that allows one to fit an adequate model and then estimate error from deviations from the model. This strategy is advocated by Bransby et al. (1988) and Riewe (1961). To illustrate the technique as well as its advantages and disadvantages, return to the example of the factorial involving herbage mass and forage cultivars. If the relationship between herbage mass and response is known to be linear, then one can design an experiment with more than two herbage masses, fit a statistical model with a *linear component* for herbage mass, and use deviations from the model to estimate experimental error. Note that exactly two different herbage mass levels are required to fit the model with a linear component for herbage mass and that each additional herbage mass level in the experiment yield one more degree of freedom for experimental error. If the response is not linear, however, for example quadratic, then at least four points are required if one wants to estimate experimental error.

Several comments about this strategy are in order. Consider three experiments, one with herbage mass levels (coded) as 1, 2, 3, and 4; a second with herbage mass levels 1 and 4; and a third with two repetitions each at levels 1 and 4. The first and third require four pastures, and the second requires two. If a model incorporating a linear response to herbage mass is fit to the data, the first and third will provide 2 df to estimate error, and the second provides none. All three provide an estimate of the linear response (regression coefficient) to increasing herbage mass level. The three regression coefficients (scaled so as to be directly comparable) will have variances proportional to 0.200, 0.267, and 0.167, respectively. There is a major gain in the sense of a reduction in variance obtained by having duplicate observations at herbage mass levels 1 and 4 rather than single observations at levels 1, 2, 3, and 4. In addition, the error estimate obtained from the deviations from regression may well be contaminated (have an upward bias) due to the true model not being linear. For example, if the true response is quadratic, then the estimated error variance will be increased by an amount due to the quadratic trend, whereas the estimate obtained from the third scheme will not be contaminated. Finally there is the often-stated comment that an experiment with more than two herbage mass levels gives one an opportunity to check on the assumption of a linear response. This may be true, but there is the risk of compromising the unbiasedness of the error term computed from deviations from the model. If one looks at how well the model fits before computing an estimate of error from deviations, then one will almost surely underestimate the experimental error if the response is linear and not quadratic.

REPEATED MEASUREMENTS

If the grazing period, *year* in the previous discussion, is broken into two seasons with data collected separately in each, then a straight forward split-plot type analysis with seasons as the split-plot factor is typically used. However, even here things are slightly more complex than they appear at first glance. The problem is that there is no real opportunity to impose a new randomization between the seasons to break up existing correlations.

As an example, consider a grazing trial with t treatments each applied to r pastures (no blocking) for m years and observations recorded for two seasons in each year. The complete formal analysis of variance is given in Table 7-4. The year × season × pasture × treatment interaction and most of the three factor interactions have been assumed negligible. This analysis does not extend directly to cases where the year is broken into more than two seasons. The difficulty is that if there are more than two periods, then observations on a pasture in adjacent time periods will have stronger correlations than observations separated by longer time intervals. In part the correlations are due to the development of the sward during the year. Some of

Table 7-4. Expected mean squares for the analysis of variance of the m year, t treatment grazing trial with each treatment replicated r times and observations in two periods.

Source†	Expected mean squares
Yr	
Ssn	$\sigma_e^2 + \sigma_{yr\times s\times p}^2 + m\sigma_{s\times p}^2 + m\sigma_{s\times p\times t}^2 + r\sigma_{yr\times s\times t}^2 + rm\sigma_{s\times t}^2 + \ldots$
Yr × ssn	$\sigma_e^2 + \sigma_{yr\times s\times p}^2 + r\sigma_{yr\times s\times t}^2 + rt\sigma_{yr\times s}^2$
Trt	$\sigma_e^2 + \sigma_{yr\times s\times p}^2 + m\sigma_{s\times p}^2 + m\sigma_{s\times p\times t}^2 + m\sigma_{yr\times p}^2 + 2m\sigma_p^2$
	$+ 2m\sigma_{p\times t}^2 + r\sigma_{yr\times s\times t}^2 + rm\sigma_{s\times t}^2 + 2r\sigma_{yr\times t}^2 + \ldots$
Pas(trt)	$\sigma_e^2 + \sigma_{yr\times s\times p}^2 + m\sigma_{s\times p}^2 + m\sigma_{s\times p\times t}^2 + m\sigma_{yr\times p}^2 + 2m\sigma_p^2 + 2m\sigma_{p\times t}^2$
Yr × trt	$\sigma_e^2 + \sigma_{yr\times s\times p}^2 + r\sigma_{yr\times s\times t}^2 + 2r\sigma_{yr\times t}^2$
Ssn × trt	$\sigma_e^2 + \sigma_{yr\times s\times p}^2 + r\sigma_{yr\times s\times t}^2 + 2m\sigma_{s\times t}^2$
Yr × ssn × trt	$\sigma_e^2 + \sigma_{yr\times s\times p}^2 + r\sigma_{yr\times s\times t}^2$
Yr × pas(trt)	$\sigma_e^2 + \sigma_{yr\times s\times p}^2 + 2m\sigma_{yr\times p}^2$
Ssn × pas(trt)	$\sigma_e^2 + \sigma_{yr\times s\times p}^2 + m\sigma_{s\times p}^2 + m\sigma_{s\times p\times t}^2$
Yr × ssn × pas(trt)	$\sigma_e^2 + \sigma_{yr\times s\times p}^2 +$
Total	

† Yr = year, ssn = season, trt = treatment, pas = pasture.

the correlations are due to growth patterns of individual animals, and some are due to such mundane things as the fill of the animals just prior to weighing. Modeling this sort of time-dependent correlation structure is extremely difficult. The problem obviously does not appear if one considers only two periods.

A simple, direct method around these difficulties is to do a simple analysis on the data for each pasture separately to estimate meaningful parameters. These estimates are then used as input into an analysis of variance of the type discussed earlier. As an example, one could compute the total digestible nutrients (TDN) produced per unit area for each pasture for the complete season, or the maximum (or minimum) TDN produced in any 2-wk period, or the ratio of the minimum to the maximum and then proceed to the overall analysis. In any case one performs a simple meaningful analysis on each pasture as a first step and then proceeds to a large-scale analysis of data from all the pastures.

COMBINING INFORMATION FROM INDEPENDENT EXPERIMENTS

Up to this point, this chapter has considered analyses of individual experiments, analyses that involve many more components of variation than there are lines in the analysis of variance. Obviously it is not possible to estimate all of the components appearing in one analysis, even for the limited models considered. The simplest way around this problem is to appeal to experience, judgement, and knowledge of the subject matter to eliminate as many components as possible. An alternative is to combine results from many separate experiments. The Petersen and Lucas (1960) paper illustrates how one can use a number of unrelated experiments to estimate parameters that are not estimable in individual experiments. They developed a general model for the components that made up the experimental error for grazing experiments. They collected observed error mean squares from statistical analyses of a large number of roughly comparable grazing trials. For each of these they used their model to express the expected values as linear functions (with known coefficients) of the unknown components of variance. They then used multiple regression (least squares) to obtain estimates of the unknown components. Although this is not the most efficient method for combining the various pieces of information (in the light of much more recent advances in statistical theory), it is still a valid and useful technique that is relatively easy to understand. It can be used to combine pieces of information on other components of variation as well.

There is a more basic problem, however, that appears to be becoming ever more pressing. There is a need to develop and use more general methods to combine information from series of experiments. As progress is being made, scientists find themselves looking for smaller and smaller effects, typically without the concomitant increases in budgets required for more sensitive experiments. More and more frequently, scientists face the problem of combining information from several limited experiments to detect small but real effects. It has been well-documented (Hedges & Olkin, 1985) that when faced with a number of small, independent studies, there is a very strong tendency to conclude that small but real effects are nonexistent.

Discussions that report only significance levels are very difficult to combine. Hedges and Olkin (1985) devote a chapter to methods for combining such studies. Simple vote-counting techniques based on fraction of studies reporting statistical significance are not adequate.

At the other extreme is the case where raw data are available from the separate experiments. The general purpose regression or least-squares programs available in most computation centers make combined analyses feasible. An assumption inherent in this sort of analysis is that the various pieces of data being combined have a common variance. Here one must be acutely aware of the appropriate model as for example discussed in the Petersen and Lucas (1960) paper. Also one is usually forced to assume that there are no other random factors in the model in addition to the residual error term. If one is not willing to assume that the errors are homogeneous and/or there happens to be a more complex random error structure, one can use a generalized least squares analysis. An example of this sort of analysis can be seen in Burns et al., (1983) and Giesbrecht and Burns (1985).

A side issue that should at least be mentioned at this point is the problem of comparability of treatments in different experiments. There is always the problem of achieving complete objectivity in research. The difficulties are not restricted to grazing experiments that often involve personal judgement at various points (Wheeler et al., 1973). One possible view is that (within reason) this should be accepted as another source of true experimental error, error in level of treatment. In fact, this may be one of the largest sources of experimental error and be one of the root causes for the common lack of agreement among pieces of research, even when some or all achieve statistical significance. Analyses of combined experiments where different experimenters—each representing slightly different views, methods, and idiosyncrasies—provide the replication that leads to the measure of experimental error, may in the long-run give us better information. After all, the purpose of the research and the publication of the interpretation is to inform others of the consequences of future actions. There are limits to the extent that given conditions and protocols can be repeated.

The majority of cases lie somewhere between the extremes of having access only to significance values (p values) and having access to the raw data. Hedges and Olkin (1985) discuss methods for combining estimates of treatment effects derived from various sources. The Petersen and Lucas (1960) paper was mentioned earlier as an example of a method for combining independent variance estimates to isolate variance components not available from individual experiments. Maximum likelihood theory provides a very efficient method for combining the pieces of information of this type to estimate components of variance. As preliminary data, we assume that we have n independent sums of squares, know the degrees of freedom associated with each, and know the expected values of the corresponding mean squares. More formally we require $\{SS_i\}$ and associated $\{df_i\}$ and assume that we can compute the $\{EMS_i\}$ that correspond to the $\{MS_i\}$ that can be computed (SS = sum of squares, EMS = expected mean squares, MS = mean squares, df = degrees of freedom). It follows that there are coefficients $\{c_{ik}\}$ and m unknown variance components $\{\sigma_k^2\}$ such that $EMS_i = \sum_k c_{ik} \sigma_k^2$. Using this information, one constructs the following nonlinear system of equations:

$$\sum_i \frac{SS_i c_{ij}}{(\sum_k c_{ik} \sigma_k^2)^2} = \sum_h \left(\sum_i \frac{c_{ih} c_{ij}}{(\sum_k c_{ik} \sigma_k^2)^2} \right) \sigma_h^2.$$

This system cannot be solved directly. However, there is a simple iterative scheme that works rather well. The trick is to replace the variance components in the denominator first by ones and then solving the linear system of equations for preliminary estimates. For the next cycle these values are inserted in the denominator and the resulting system again solved. This process of repeatedly substituting existing solutions and resolving is continued until there are no further changes. Note that if there are m variance components, then a system of m linear equations in m unknowns must be solved each cycle. Experience shows that the system tends to converge rapidly, usually in two or three cycles.

As a final point, it is important that we develop and use adequate methods to combine information from series of experiments. In addition there is a need to put more emphasis on documenting results of experiments with sufficient detail to make the task of combining results as efficient as possible. This appears to be the only realistic way around the logistical and financial constraints problem that forces compromised in the sizes of grazing experiments. The individual researcher's rights and responsibilities to conduct appropriate grazing trials are protected. The researcher can make what he/she considers reasonable assumptions in his/her interpretations thereof. Yet, it provides a framework that allows a more coherent body of information to develop. The individual experimenter may have to make all sorts of restrictive assumptions about certain sources of error being zero or at least negligible to interpret his/her results. Combined experiments may give an opportunity to check some of these assumptions.

REFERENCES

Bransby, D.I., B.E. Conrad, H.M. Dicks, and J.W. Drane. 1988. Justification for grazing intensity experiments: Analysing and interpreting grazing data. J. Range Manage. 41:274–279.

Burns, J.C., F.G. Giesbrecht, R.W. Harvey, and A.C. Linnerud. 1983. Central Appalachian hill land pasture evaluation using cows and calves. I. Ordinary and generalized least squares analysis for an unbalanced grazing experiment. Agron. J. 75:865–871.

Fisher, R.A. 1953. The design of experiments. Hafner Publishing Co., New York.

Giesbrecht, F.G., and J.C. Burns. 1985. Two-stage analysis based on a mixed model: Large sample asymptotic theory and small-sample simulation results. Biometrics 41:477–486.

Hedges, L.V., and I. Olkin. 1985. Statistical methods for meta-analysis. Academic Press, New York.

Mott, G.O., and H.L. Lucas. 1952. The design, conduct and interpretation of grazing trials on cultivated and improved pastures. p. 1380–1385. *In* Proc. 6th Int. Grassl. Congr., State College, PA. 17-23 August. Pennsylvania State Univ., State College, PA.

Petersen, R.G., and H.L. Lucas. 1960. Experimental errors in grazing trials. p. 747–750. *In* Proc. 8th Int. Grassl. Congr., Reading, England. 11-21 July. Alden Press, Oxford, England.

Riewe, M.E. 1961. Use of the relationship of stocking rate to gain of cattle in an experimental design for grazing trials. Agron. J. 53:309–313.

Steel, R.G.D., and J.H. Torrie. 1980. Principles and procedures of statistics. McGraw-Hill, New York.

Wheeler, J.L., J.C. Burns, R.D. Mochrie, and H.D. Gross. 1973. The choice of fixed or variable stocking rates in grazing experiments. Exp. Agric. 9:289–302.

8 Time Series, Dynamic Models, and Adaptive Sequential Decisions in Grazing Research

Donald A. Jameson
Colorado State University
Fort Collins, Colorado

ABSTRACT

An objective for developing new designs for grazing research is to reduce sampling cost or to reduce experimental errors. Predictions from a simple time series model modified by external effects such as precipitation and forage harvested can be combined with sample results in a linear filter to achieve these goals. These improved estimates and predictions can be used adaptively to guide managers toward an optimal stocking rate under certain well-specified conditions. Economic analyses usually assume that the conditions for incremental convergence to an optimum exist; ecologists frequently cite concepts of hysteresis and discontinuity to indicate that incremental convergence to an optimal stocking rate is not possible.

For any research design, goals include achieving a suitably small experimental or sampling error and sampling at a suitably low cost. New measurement techniques or new approaches to experimental design usually have one or both of these goals. In this chapter we address the question of reducing statistical error for a given cost, or of reducing cost for a given level of statistical error, by combining predictions with sampling results. We also examine whether or not such procedures can be used to answer questions such as optimal stocking rates.

Most experimental and sampling procedures, especially when designed by an investigator with a background in Fisherian statistics, assume that an experiment conducted at a given time represents an independent estimate of the condition or state of the system under study for that instant of time. At a later date, another experiment is assumed to represent another independent estimate. In grazing research, several such independent results are often used to make recommendations in stocking rate for yet another time period. If two samples of a system are independent in time, what can we say about the condition or state of the system between Time 1 and Time 2? Unless we

---

Copyright © 1989 Crop Science Society of America and American Society of Agronomy, 677 S. Segoe Rd., Madison, WI 53711, USA. *Grazing Research: Design, Methodology, and Analysis*, CSSA Special Publication no. 16.

include a statement of how the system undergoes a transition between these two times, we cannot make any interpretation of the intervening period, nor can we make any predictions about a future time. If we do make some statement of the interval between Time 1 and Time 2, we must have in mind some dynamic transition model. The purpose of this chapter will be to examine the use of dynamic transition models to improve the statistical and managerial interpretation of grazing research.

TIME SERIES

The basic linear regression model follows:

$$y = a + bx + e$$

where y is a dependent variable, x is an independent variable, a is the intercept or value of y when $x = 0$, and b is the slope of the regression line. The symbol e indicates a residual error that is supposed to be normally distributed with zero mean, and the variance of e is that portion of the variance of y not explained by the regression line. The variance of e is minimized by the choice of a and b. The model

$$y = bx$$

is interpreted as a linear equation with zero intercept.

A simple time series regression results when the numerical values of the dependent variable y are the numerical values for the system at one time step later than values of the system indicated by the independent variable x, that is

$$x_{k+1} = bx_k$$

This equation is a simple linear regression, calculated numerically like the previous example $y = bx$.

If we plot the results of $x_{k+1} = bx_k$ over time k, $b = 1$ will produce a straight line with constant value. Values of $0 < b < 1$ will produce a line that asymptotically declines to zero, and values of $b > 1$ will produce a line that increases over time. No biological system is likely to follow this simple example for a long period. In the same way, no biological system really follows the linear regression $y = a + bx$, but we might still be willing to use this simplifying assumption for convenience. We can, however, readily add flexibility to a time series by choosing a different value of b for different time intervals

$$x_{k+1} = b_1 x_k \text{ for time periods 1 to 10, } b_1 > 1$$

$$x_{k+1} = b_2 x_k \text{ for time periods 11 to 20, } b_2 = 1$$

$$x_{k+1} = b_3 x_k \text{ for time periods 21 to 30, } 0 < b_3 < 1$$

TIME SERIES, DYNAMIC MODELS, & ADAPTIVE SEQUENTIAL DECISIONS

In this example, the coefficient b is in effect dependent on time, but the system is nevertheless linear in the parameters for any time period k to $k + 1$. Thus, we can use all of the statistical procedures appropriate to linear dynamic systems for any series of intervals for which b_k can be mathematically specified.

We could easily expand the regression equation to a multiple regression indicated by

$$y = b_1 x_1 + b_2 x_2$$

in traditional regression notation. In time series or state-space notation, the equation is

$$x_{k+1} = b_1 x_k + b_2 u_k$$

where u indicates some external variable that is independent of the time series of x. A simple example in grazing research might be a series of forage weights indicated by x and precipitation indicated by u; in these cases we would expect $b_2 u_k$ to be positive, and a single value of $0 < b_1 < 1$ to be appropriate for several to many time intervals. In other words, water makes the forage grow; without water the forage will eventually decline to zero.

In the time series literature (Box & Jenkins, 1976; Chatfield, 1984), the coefficient b of conventional regression is commonly indicated by the Greek letter α. In the literature on linear dynamic systems, b_1 is commonly indicated by the Greek letter ϕ, b_2 by b, and the letter w is often used for random error instead of the letter e. In the scalar case (a single dependent variable), the meaning of the coefficients are the same regardless of the symbol used.

LINEAR DYNAMIC SYSTEMS AND THE KALMAN FILTER

A time series is a familiar linear regression with somewhat different symbols

$$x_{k+1} = \phi x_k + w_k, \ w \text{ is } N(0, q)$$

where x is the variable of the system whose dynamics are indicated by the equation, ϕ is the time series regression coefficient or state transition multiplier, and w_k is a random error from a population that is normally distributed with zero mean and variance q. Recall that q is that portion of the variance of x_{k+1} not explained by the regression. The system in this equation is free, in that it is not controlled by any external events. For a controlled system,

$$x_{k+1} = \phi x_k + b u_k + w_k$$

where u_k indicates some external control (such as precipitation) as in the example above.

Although we may assume that a dynamic grazing system follows this equation, we are not able to determine with certainty that the regression model actually occurs in nature. The interpretation is restricted by sampling error. For a sample at any time period, an observation is

$$y_{k+1} = x_{k+1} + v_{k+1}, \quad v \text{ is } N(0, r)$$

where y is the observed value, x is the true state of the system, and v is a random error of observation. For unbiased samples, the random variable v_{k+1} comes from a population that is normally distributed with zero mean and variance r (biologists would call r the square of the standard error of the mean, and it depends on the sample size; in the engineering literature r is called a variance). Thus, even though the sample is unbiased, the state of the system is uncertain because of the random sampling effects indicated by v and r.

Assume that we have collected a chronological series of data and developed the time series equation

$$x_{k+1} = \phi x_k + w_k$$

Depending on the variance q of w, we should be able to determine an expected value for x_{k+1} if we have information about x_k. Even if the variance q is small, the sample value y_{k+1} may be different than the predicted value x_{k+1}. If the sample value were perfect (that is, the variance r is zero), we could just accept the sample value and ignore the prediction. Assume, however, that variance q is small and variance r is large; in this case the prediction is better than the sample. Should we then accept the prediction and ignore the sample?

Fortunately, we are able to combine the sample value and the predicted value and produce a weighted average that has a lower variance than either the prediction or the sample. Is this a valid procedure? Consider what we do when we have two subsamples that we went to combine; we could simply take $(y_1 + y_2)/2$ to get an arithmetic mean. We get the same result with

$$0.5y_1 + 0.5y_2$$

where each of the subsamples is weighted by a factor of 0.5. We may not have considered that an unbiased average of two subsamples assumes that each subsample has the same standard error of the mean. To combine subsamples with unequal standard errors, we should devise weighting factors so that a subsample with a high standard error has a small weighting factor and a subsample with a low standard error has a large weighting factor. The appropriate weighting factor g_1 for y_1 is

$$g_1 = r_2/(r_1 + r_2)$$

and the weighting factor g_2 for y_2 is

$$g_2 = r_1/(r_1 + r_2)$$

Thus $g_1 + g_2 = 1$, or $g_2 = (1 - g_1)$, just as in the example above when $g_1 = g_2 = 0.5$.

Therefore, if we want to combine a prediction with a sample, we could do so if we knew the variance of the prediction and of the sample. We certainly can determine the variance r (square of the standard error of the mean) of a sample, and the variance q (from the sum of squares independent of regression; see Cochran, 1977). To find the variance of a prediction, assume a grazing system was sampled at time k, and thus our knowledge of the system is indicated by variance r. If we make a prediction to time $k + 1$, we must add the variance q to indicate the variance appropriate for the prediction. Thus,

$$p = r + q$$

where p is the variance appropriate for the prediction of x at time $k + 1$ if r and q are independent of each other. We can calculate the weighting factor g by

$$g = p/(p + r)$$

where g is the weighting factor for the sample, and $(1 - g)$ is the weighting factor for the prediction. Thus,

Combined or best estimate at time $k + 1$

$$= g \text{ (Sample)} + (1 - g)(\text{Prediction})$$

The prediction for time $k + 1$ uses only information available at time k; thus, we indicate the prediction by $x_{k+1|k}$, which is read as x_{k+1} given k. When we update the information at time $k + 1$ and combine the sample and prediction, then we have $x_{k+1|k+1}$, which is read as x_{k+1} given $k + 1$. Thus, the equation for the updated best estimate is

$$x_{k+1|k+1} = gy + (1 - g) x_{k+1|k}.$$

With this information, we can then make a prediction for time $k + 2$ based on knowledge of the state transition multiplier ϕ and the best available information for time $k + 1$. However, to evaluate the prediction, we need to calculate a variance of the combined information at time $k + 1$. When we have incorporated the results of a new sample, the variance p at time $k + 1$ will be less than either the variance q or the variance r; thus, we have achieved a reduction in statistical error only by combining two sources of information. In fact, we have combined knowledge of the past dynamic behavior of the system with knowledge of the current sample. If we indicate the variance of the prediction as $p_{k+1|k}$, then the variance appropriate for $x_{k+1|k+1}$ is

$$p_{k+1|k+1} = g^2 r + (1 - g)^2 p_{k+1|k}$$

The exact development of this latter equation is outlined in Jazwinski (1970), Gelb (1974), Maybeck (1979), and Anderson and Moore (1979), but the equation for $p_{k+1|k+1}$ is similar to the previous equation for $x_{k+1|k+1}$. The variance $p_{k+1|k+1}$ will be less than either r or q.

The above equations can be summarized by the Kalman filter, which follows

Identify the time series or linear dynamic system as

$$x_{k+1} = \phi x_k + w_k, \; w \text{ is } N(0, q)$$

the measurements as

$$y_{k+1} = x_{k+1} + v_{k+1}, \; v \text{ is } N(0, r)$$

and the prediction as

$$x_{k+1|k} = \phi x_{k|k}$$

The variance of the prediction is

$$p_{k+1|k} = \phi^2 p_{k|k} + q$$

and the weighting factor for the measurement is

$$g = p_{k+1|k}/(p_{k+1|k} + r)$$

The best estimate at time $k + 1$ is

$$x_{k+1|k+1} = g y_{k+1} + (1 - g) x_{k+1|k}$$

and the variance of the best estimate is

$$p_{k+1|k+1} = g^2 r + (1 - g)^2 p_{k+1|k}$$

At time $k = 0$, take a sample and assume $p_{k|k} = r$. Solve for $x_{k+1|k+1}$ and $p_{k+1|k+1}$. If no measurement is taken, however, at a time step, $p_{k+1|k+1} = p_{k+1|k}$ and $x_{k+1|k+1} = x_{k+1|k}$. For the next time step, in either case, set $p_{k|k} = p_{k+1|k+1}$, $x_{k|k} = x_{k+1|k+1}$, and repeat the calculations.

Use of Kalman filter approaches in efficient sampling design is presented in Carande and Jameson (1986), Jameson (1986a), and Mosetti (1988). Further details and applications to multivariable models can be found in Jameson (1988). Use of multivariable models allows combination of variables such as direct and indirect measurements, but the relationship to simple regression is more circuitous than for the examples used in this chapter.

COMBINATION WITH EXTERNAL VARIABLES

We earlier discussed the possibility of including an external variable such as precipitation in the time series. We might expect to find that the added variable of precipitation reduces the magnitude of the prediction variance q, in addition to helping with an updated estimate of x. To predict precipitation, unless we develop some significant breakthrough in weather forecasting, the historical average precipitation for the interval k to $k + 1$ may be used. The linear dynamic equation for the state variable x in this case is

$$x_{k+1|k} = \phi x_{k|k} + bu_k$$

where u is the precipitation, and ϕ and b are regression coefficients found by the linear multiple regression equation using x_{k+1} as the dependent variable and both x_k and actual precipitation u_k as independent variables.

Although we may predict using long-term average precipitation, the actual precipitation will vary. Therefore, we can repeat the calculations using actual precipitation when the data become available; this should improve the accuracy of estimates of available forage for management decisions. However, if the forage model values begin to drift apart from the actual values, or the estimates for variance $p_{k+1|k+1} = p_{k+1|k}$ become too large when no measurements are taken, then the forage model values must be corrected by combining a forage sample with the forage prediction using appropriate weighting factors as calculated above.

INTRODUCTION OF MANAGEMENT CONTROLS

The example above using a simple time series, modified by precipitation effects, to predict some state of a grazing system such as available forage, can be very useful. In a grazing management study, we might also want to expand the equation to include some management activity such as forage harvesting by livestock. The resultant equation

$$x_{k+1} = \phi x_k + b_1 u_1 + b_2 u_2$$

indicates that the predicted value for x (available forage in this example) depends on the current value for x, the expected or predicted value of precipitation (u_1) and some management variable u_2.

Methods of expanding linear dynamic equations into an optimization model such as linear programming or dynamic programming are given in Jameson and Bartlett (1987). In grazing research, however, it might be preferable to use a model to investigate system behavior by simulation. In any event, we can include the forage harvested by grazing ($b_2 u_2$) as a variable in the equation; u_2 might be the number of grazing animals and b_2 the forage removed per animal. It may be more satisfactory to determine $b_2 u_2$ from research on animal intake rather than from regression analysis with pasture data, however. Because forage is removed, the numeric value of b_2u_2 in the forage weight equation should be negative. The combination of predicted forage and sampled forage data proceeds as outlined above except that an alternative method of calculating the variance q must be used (Jameson, 1985).

A similar and concurrent model (without precipitation) could be designed for animal weight gains; the value of b_2u_2 in the animal model must be positive. A challenge in developing an animal model is to find an appropriate relationship between forage available and forage intake. Separating the forage variable into time periods of positive vs. negative growth increments might be a suitable way to distinguish between periods of high vs. low nutrient value, and thus allow forage available to be more closely related to animal intake and weight gains than would using total forage biomass alone.

In a linked plant-animal model, we expect that if the product b_2u_2 were increased (more forage harvested), x_{k+1} would decrease, but total animal weight gain would increase (at least within limits). A corresponding decrease in b_2u_2 (less forage harvested) results in a similar increase in x_{k+1} and less total animal weight gain. If these conditions hold, we can then use such an equation to guide grazing management to satisfactory levels of forage production and animal weight gain by making a series of predictions of available forage and adjustments in numbers of grazing animals. Thus, the model provides a basis for adaptive management of stocking rate. I argued that an adaptive procedure is inherently superior to attempts to make fixed recommendations of stocking rate (in a highly stochastic environment) based on traditional replicated experimental designs (Jameson, 1986b). The challenge to research with an adaptive approach is to develop suitable equations and measurement specifications for use in adaptive management.

A simple predictive model similar to that outlined here can be used for incremental convergence to an optimal stocking rate strategy if some relatively simple conditions exist (adapted from Bar-Shalom & Tse, 1976; Casti, 1980):

1. The ecological effects of a series of increases in stocking rate (u_2) can be reversed by a corresponding series of decreases
2. The economic loss of a given amount of overstocking is the same as the economic loss of a like amount of understocking
3. The random variables are normally distributed, or at least the variances are independent and additive as indicated above

Let us focus on the ecological effects noted in the first of these three conditions. To study the appropriateness of this condition, we can conduct an experiment with a series of increases of u_2, that is, to increase forage removed (b_2u_2) over time on a pasture with a past history of light stocking. With this experiment we can produce a trace of available forage x over stocking rate u_2. Separately, we can conduct another experiment with a series of decreases of u_2, that is, to decrease forage removed (b_2u_2) over time on a pasture with a past history of heavy stocking. For the simplest cases, these experiments would result in a trace of x as u_2 increases that is the same as the trace of x over u_2 as u_2 decreases. This means that past stocking rate has no influence on current forage production. To complete the analysis, we need data to relate animal gains to forage available. In these cases, we can use the model to converge to a stocking rate that lies between the economically rational bounds of maximum total weight gain and maximum gain per animal.

Economic cost and benefit data are required for a precise determination of an economic optimum, but because of constantly changing costs, livestock prices, and forage production, it is likely that the optimal stocking rate will never be exactly achieved (Parsch & Torell, 1989). Because of these uncertainties, maintaining a grazing system between the rational bounds will be a challenge in itself. Greenwood (1986) suggested that management strive to achieve maximum gain per animal during periods of abundant forage and maximum total gain during periods of deficient forage; some experimental data indicate that this strategy would give a nearly constant stocking rate that is always within the rational bounds (Marten & Jordan, 1972).

On the other hand, if increasing u_2 results in a different trace than decreasing u_2, the system has the properties of hysteresis and discontinuity, and the model as developed above cannot be used to find an optimal stocking rate. Most economic analyses (for example, Workman, 1986) assume that optimal stocking rates can be found through convergence, that is, by incrementally increasing or decreasing the stocking rate. Convergence is possible or even likely in pastures that can be managed as though they contain a single plant species, but pastures with two or more plant species are a different matter. Although I do not know of an experiment where the question has been explicitly studied, many ecological concepts (Walker et al., 1981; Johnson & Parsons, 1985; Loehle, 1985; Jameson, 1987) suggest that optimal stocking in plant communities that contain competing plant species with different preferences to grazing animals cannot be achieved by convergence. In these cases, the effects of competition between the plant taxa as modified by grazing must be explicitly expressed.

Additional analyses might be required to determine the appropriateness of the last two conditions required for incremental convergence, but on the surface these conditions seem to be less important for grazing management than the hysteresis/competition effect that violates the first condition.

CONCLUSIONS

Predictions resulting from a time series model, possibly expanded to include precipitation and grazing effects, can be combined with forage samples to improve statistical reliability of forage information. The procedures are straightforward and are well-documented in the literature. Use of time series models of both forage and animals, linked through the common variable of forage removed or harvested, can in some cases be used to converge to an optimal stocking rate that lies between the rational bounds of maximum gain per animal and maximum total animal gain. In cases with competing plant taxa that have different preferences to grazing animals, it appears that identifying optimal stocking rates through the usual economic analyses and incremental convergence is not possible without an explicit analysis of plant competition effects.

The analytical and management methods presented here depend on a time series or linear dynamic model that begins with some initial estimate of available forage and is periodically updated with precipitation measurements. Samples to determine forage weight may be required when the forage weights estimated by the model begin to drift apart from the actual values, or the system variance becomes too high. In addition, the following specific items of biological research are needed to complete the model:

1. Research to determine the quantitative relationship of forage availability to animal intake and weight gains
2. Research to determine the effect of grazing on interspecific or intervarietal plant competition
3. Research to determine whether changing a stocking rate from light to heavy will result in the same trace of forage available plotted over stocking rate as changing stocking rate from heavy to light

REFERENCES

Anderson, B.D.O., and J.B. Moore. 1979. Optimal filtering. Prentice-Hall, Englewood Cliffs, NJ.
Bar-Shalom, Y., and E. Tse. 1976. Concepts and methods in stochastic control. Control Dyn. Syst. 12:99–172.
Box, G.E.P., and G.M. Jenkins. 1976. Time series analysis: Forecasting and control. Holden-Day, San Francisco.
Carande, V., and D.A. Jameson. 1986. Combination of weight estimates with clipped sample data. J. Range Manage. 39:88–89.
Casti, J.L. 1980. Bifurcations, catastrophes and optimal control. IEEE Trans. Autom. Control AC-25:1008–1011.
Chatfield, C. 1984. The analysis of time series. Chapman and Hall, New York.
Cochran, W.G. 1977. Sampling techniques. John Wiley & Sons, New York.
Gelb, A. 1974. Applied optimal estimation. MIT Press, Cambridge.
Greenwood, G.B. 1986. Does sahelian pasture development include range management? Rangelands 8:259–264.
Jameson, D.A. 1985. A priori analysis of allowable interval between measurements as a test of model validity. Appl. Math. Computation 17:93–105.
Jameson, D.A. 1986a. Sampling intensity for monitoring of environmental systems. Appl. Math. Computation 18:71–75.
Jameson, D.A. 1986b. What shall we do about grazing systems research? Rangelands 8:178–179.
Jameson, D.A. 1987. Climax or alternative steady states in woodland ecology. p. 9–13. In R.L. Everett (compiler) Proc.—Pinyon-Juniper Conf., Reno, NV. 13-16 Jan. 1986. Rep. GTR INT-216. USDA Forest Service Intermountain Res. Stn., Ogden, UT.
Jameson, D.A. 1988. Modelling rangeland ecosystems for monitoring and adaptive management. p. 189–221. In P.T. Tueller (ed.) Vegetation science applications for rangeland analysis and management. Kluwer Publ., Dordrecht, the Netherlands.
Jameson, D.A., and E.T. Bartlett. 1987. Selection of optimal management strategies based on stochastic dynamic ecological models. Ecol. Modell. 36:5–13.
Jazwinski, A.H. 1970. Stochastic processes and filtering theory. Academic Press, New York.
Johnson, I.R., and A.J. Parsons. 1985. A theoretical analysis of grass growth under grazing. J. Theor. Biol. 112:345–367.

Loehle, C. 1985. Optimal stocking for semi-desert range: A catastrophe theory model. Ecol. Modell. 27:285-297.

Marten, G.C., and R.M. Jordan. 1972. Put-and-take vs. fixed stocking for defining three grazing levels by lambs on alfalfa-orchardgrass. Agron. J. 64:69-72.

Maybeck, P.S. 1979. Stochastic models, estimation, and control. Vol. 1. Academic Press, New York.

Mosetti, R. 1988. Optimal strategies for monitoring a variable subjected to random changes. Appl. Math. Computation 25:137-143.

Parsch, L.D., and L.A. Torell. 1989. Economic considerations in grazing research. p. 109-125. *In* G.C. Marten (ed.) Grazing research: Design, methodology, and analysis. CSSA Spec. Publ. 16. ASA, CSSA, Madison, WI (Chapter 9 of this publication).

Walker, B.H., D. Ludwig, C.S. Holling, and R.M. Peterman. 1981. Stability of semi-arid savanna grazing systems. J. Ecol. 69:473-498.

Workman, J.P. 1986. Range economics. MacMillan, New York.

9 Economic Considerations in Grazing Research

Lucas D. Parsch
University of Arkansas
Fayetteville, Arkansas

L. Allen Torell
New Mexico State University
Las Cruces, New Mexico

ABSTRACT

The experimental design of grazing trials affects the ability to provide information relevant for economic analysis. A key management issue—optimal stocking rate on range or improved pasture—is analyzed from the perspective of both profitability and risk-return tradeoffs. To identify the most profitable grazing system, the experimental design of stocking rate studies must include a broad range of treatment levels (stocking rates) so that weight gain response can be determined. To assess the impact of weather risk, data must be collected in both good and bad years so that probabilistic estimates of weight gains and economic returns can be determined. Experiments show that (i) stocking rates that maximize average daily gain or weight gain per unit area do not necessarily maximize profits, and (ii) high stocking rates increase the exposure of the producer to risk.

The objective of researchers who conduct grazing studies is to advance disciplinary knowledge by enhancing understanding of the response of crops and/or livestock in a range or pasture ecosystem. Nevertheless, the ultimate end-user of grazing study research is the producer, who must decide how to economically manage resources. The research output of grazing studies need not be devoid of managerial information for use in a decision support mode by producers. However, if research is to have managerial implications as well as advance disciplinary knowledge, researchers must be cognizant of producer economic objectives, and they may also have to design grazing studies to include treatments that reflect the relevant managerial options that the producer faces.

This study examines how incorporation of managerial economic objectives into grazing studies may influence the design of research. Two producer

---

Copyright © 1989 Crop Science Society of America and American Society of Agronomy, 677 S. Segoe Rd., Madison, WI 53711, USA. *Grazing Research: Design, Methodology, and Analysis*, CSSA Special Publication no. 16.

economic objectives that seek to identify the optimal stocking rate are explored. These objectives are (i) to maximize net returns or profit per unit area of pasture; and (ii) to accommodate the tradeoff between risk and economic returns.

The issue of economic objectives is important because the various disciplines that conduct research on grazing systems have typically used different variables to measure system performance. Animal scientists have frequently concentrated on developing grazing systems that maximize average daily gain (ADG) or other per-head animal performance. Range scientists and agronomists have typically emphasized the productivity of the forage stand and defined an optimum stocking rate based on sustained forage yield or animal product per unit area. In general, neither of these two goals—growing the biggest steer or producing the most beef per hectare without deterioration of the grass stand—will maximize profits. In the first case, stocking rates are too low and in the second case, they are too high, given typical price/cost relationships.

In this study, the economic principles involved in identifying the optimal stocking rate are illustrated using data collected in a grazing trial (Sims et al., 1976). The issues of research design and the important data required to adequately quantify livestock and forage response for economic evaluation are also discussed. Although yearling steers (*Bos* sp.) and season-long grazing are illustrated, the economic principles examined have general application to all categories of grazing systems and livestock types.

STOCKING RATES AND NET RETURNS PER HECTARE

Published results of grazing studies show general agreement on two issues: (i) low stocking rates result in the highest weight gain or average daily gain (ADG) *per animal* (Chapman et al., 1972; Guerrero et al., 1984); and (ii) an increase in the stocking rate to medium or high levels results in the maximum weight gain *per unit area of pasture* (Hull et al., 1965; Bement, 1969; Neville & McCormick, 1976; Adjei et al., 1980; Willms et al., 1986). Some researchers have modeled response surfaces of the relationships in (i) and/or (ii) to enable determination of the stocking rate, which results in either maximum weight gain per head or maximum weight gain per hectare (Harlan, 1958; Riewe, 1961; Riewe et al., 1963; Peterson et al., 1965; Hart, 1972; Jones & Sandland, 1974; Hart, 1978; Conrad et al., 1981).

Investigations into the economics of stocking rates have likewise resulted in general agreement that neither the highest weight gain per animal nor maximum weight gain per unit area of pasture renders maximum net returns, but rather that the economic optimum for a typical set of price relationships exists at an intermediate stocking rate (Hildreth & Riewe, 1963; Bement, 1969; Hart, 1972; Riewe, 1981; Russell et al., 1981; Workman & Fowler, 1986; Hart et al., 1988; Torell & Hart, 1988).

The principles and conditions necessary to determine the profit maximizing stocking rate are well-defined in the production economics literature. Economic analysis to determine the stocking rate that renders maximum net returns per hectare stems from the perspective that pasture or rangeland is the fixed and most limiting resource, whereas the livestock are variable resources. Livestock are variable in the sense that the number of steers per hectare—i.e., the stocking rate (SR)—will be set at a level that results in optimal weight gain and maximum net returns per hectare.

Biophysical Response

Underlying this economic evaluation is the basic concept of biophysical input/output response, i.e., the production function, which relates the weight gain per hectare (WTGNHA, kg ha^{-1}) in Eq. [1] as a continuous function g of the number of steers (SR, head ha^{-1}) placed onto the pasture:

$$\text{WTGNHA} = g(\text{SR}) \qquad [1]$$

Unquestionably, the weight gain per hectare is coincidentally a function of available forage. The major point, however, is that from the managerial perspective the stocking rate decision is made at the beginning of the grazing season based on the producer's expectation of forage availability. Hence, as grazing system manager, the producer follows a decision rule that sets the *input* (SR) at a level expected to result in the optimal *output* level. From an experimental design perspective, this suggests that SR—as a managerial input—is a treatment variable and not a response.[1]

The WTGNHA response in Eq. [1] is typically derived from experimental field data on weight gain per head (WTGNHD, kg head^{-1}). If the seasonal WTGNHD is expressed as a continuous function f of SR as in Eq. [2], then WTGNHA is the product of WTGNHD and SR as in Eq. [3]:

$$\text{WTGNHD} = f(\text{SR}) \qquad [2]$$

$$\text{WTGNHA} = \text{WTGNHD} \times \text{SR} = f(\text{SR}) \times \text{SR} \qquad [3]$$

The WTGNHA response can also be derived from average daily gain (ADG). If ADG (kg head^{-1} d^{-1}) is expressed as a continuous function h of SR as in Eq. [4], then the seasonal WTGNHD response (Eq. [5]) is found by multiplying Eq. [4] by the length of the grazing period (t) in days. The resulting seasonal response of WTGNHA is then expressed as in Eq. [6]:

$$\text{ADG} = h(\text{SR}) \qquad [4]$$

$$\text{WTGNHD} = t \times \text{ADG} = t \times h(\text{SR}) \qquad [5]$$

$$\text{WTGNHA} = \text{WTGNHD} \times \text{SR} = [t \times h(\text{SR})] \times \text{SR} \qquad [6]$$

---

[1] In this discussion, *treatment* refers to a specified SR level (as in a one-factor experiment) to focus on the stocking rate decision as the key managerial variable.

Many researchers (Peterson et al., 1965; Hildreth & Riewe, 1963; Jones & Sandland, 1974; Riewe, 1981; Hart et al., 1988; Bransby et al., 1988) have found the functional relationship h in Eq. [4] to be best described by a linear equation within the area of economic interest, as shown in Eq. [4a] below, where a and b are estimated parameters for the intercept and slope, respectively:

$$\text{ADG} = a - b \times \text{SR} \qquad [4a]$$

Given a linear ADG response, it follows that the WTGNHD response is also a linear function of SR (Eq. [5a]), whereas WTGNHA is of the quadratic form (Eq. [6a]) and hence, curvilinear:

$$\text{WTGNHD} = t \times \text{ADG} = (t \times a) - (t \times b \times \text{SR}) \qquad [5a]$$

$$\text{WTGNHA} = \text{WTGNHD} \times \text{SR}$$
$$= (t \times a \times \text{SR}) - (t \times b \times \text{SR}^2) \qquad [6a]$$

The most noteworthy issue in comparing the response surfaces in Eq. [4a], [5a], and [6a] is the tradeoff that occurs between the *per head* and *per hectare* measures of animal performance. The largest ADG and WTGNHD occur at low stocking rates where there is little competition for desirable forage species. Because of greater forage availability, much of the forage matures and goes unused, resulting in lower weight gains per unit of land than the maximum amount possible. As SR is increased, more intense competition for forage causes individual WTGNHD and ADG to decline, resulting in smaller steers being sold. Nevertheless, the additional animals at higher stocking rates add to total beef production per hectare until maximum WTGNHA is achieved.

These tradeoffs are clearly depicted in Fig. 9-1 and 9-2, which are adapted from a long-term grazing study reported by Sims et al. (1976). This Colorado study was conducted using light, moderate, and heavy grazing intensities for yearling steers on heterogenous range species consisting primarily of blue grama (*Bouteloua gracilia* H.B.K. Lag.), prairie sandreed [*Calamovilfa longifolia* (Hook) Scribn.] and needle-and-thread (*Stipa comata* Trin. & Rupr.). Both livestock and forage response to grazing treatments were measured.

Figure 9-1 shows a linear, negatively sloped ADG function of the form found in Eq. [4a] statistically fitted to the Sims et al. (1976) data. The figure uses the letters L, M, and H to designate how Sims et al. classified the data into light, moderate, and heavy SR treatments. In Fig. 9-2, this same ADG function and data are converted to WTGNHD for a 150-d grazing period using Eq. [5a]. Note that this linear, negatively sloped WTGNHD function has been mathematically extrapolated well beyond the Sims et al. data set to 2.50 head ha^{-1} where predicted WTGNHD is approximately 47 kg head^{-1} for the grazing season. Much higher weight gains of approximately 110 kg head^{-1} are predicted at the lowest experimental SR levels of 0.22 head ha^{-1}.

ECONOMIC CONSIDERATIONS

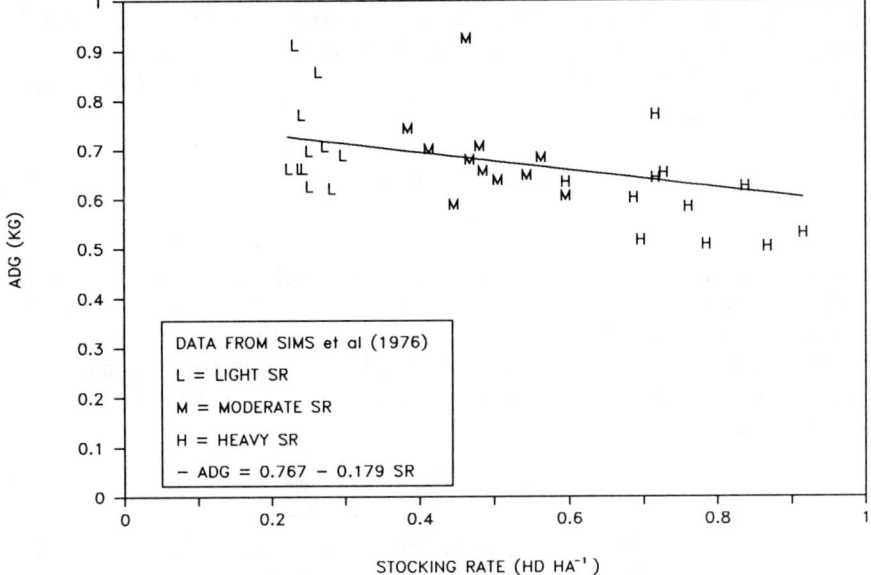

Fig. 9-1. Estimated ADG response of yearling steers to three stocking rates of range grasses.

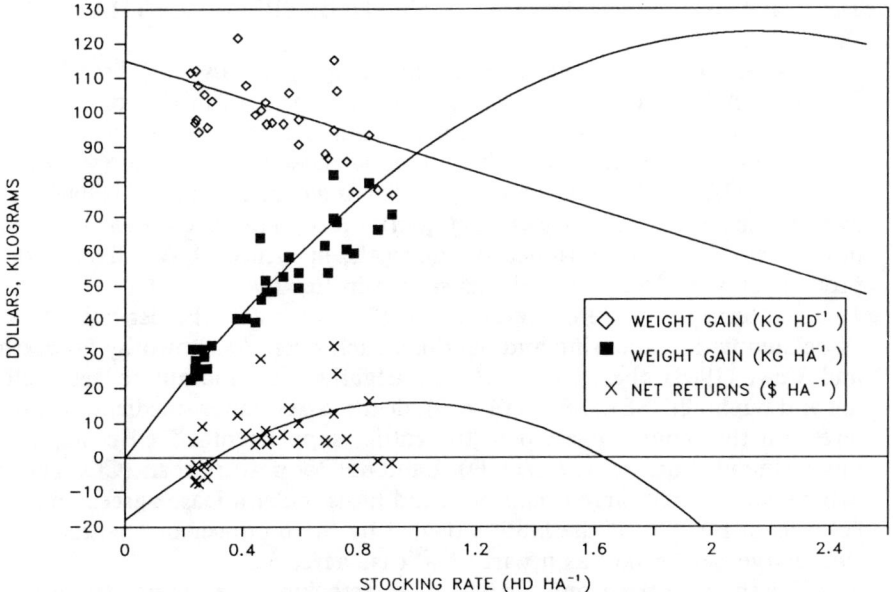

Fig. 9-2. Predicted response of WTGNHD, WTGNHA, and NR to steer stocking rates of range grasses.

Figure 9-2 also depicts the data and implicit response surface for WTGNHA using Eq. [6a], which likewise has been extrapolated beyond the experimental data. Over the broad range of SR, predicted weight gains increase quadratically and reach a maximum 123 kg ha^{-1} at 2.14 steers ha^{-1} despite the smaller per-head weight gains (58 kg ha^{-1}).

Economic Response

The economic performance measure, net returns per hectare of land (NR), is fundamentally a functional relationship of SR in that the function f(SR) of Eq. [2]—and implicitly, g(SR) of Eq. [1]—is embedded in the NR function. In its most basic form, NR ($ ha^{-1}) is simply the sum of gross receipts (GR) from the sale of the steers ($ ha^{-1}) minus total costs (TC) associated with their purchase and maintenance in addition to land or pasture costs ($ ha^{-1}) as in Eq. [7]:

$$NR = GR - TC \qquad [7]$$

Given the stocker purchase weight (WTBUY, kg head^{-1}), stocker purchase and selling prices (PSELL and PBUY, $ kg^{-1}), per head livestock costs (COSTHD, $ head^{-1}), and land or pasture costs (COSTHA, $ ha^{-1}), Eq. [7] can be expanded as follows:

$$GR = PSELL \times [(WTBUY + WTGNHD) \times SR] \qquad [7a]$$

$$TC = \{[(PBUY \times WTBUY) + COSTHD] \times SR\} + COSTHA \qquad [7b]$$

Inspection of the terms in Eq. [7a] and [7b] reveals why the NR Eq. [7] is essentially an economic response surface that is a function of SR. The term in square brackets in Eq. [7a] is the total per-hectare sales weight (kg ha^{-1}), which is the product of the individual steer selling weight (WTBUY + WTGNHD) and SR. Although WTBUY is a constant and WTGNHD is a linear function of SR (see again Eq. [5a]), the total per-hectare sales weight in brackets will vary curvilinearly due to the influence of WTGNHA (the product of WTGNHD and SR) as shown in Eq. [6a].

The term in square brackets in Eq. [7b] represents the per-head costs associated with purchasing and handling each steer. As shown by Hildreth and Riewe (1963), these costs include a negative price margin between selling and buying [(PSELL-PBUY) < 0], death losses, labor, medications, interest on the money invested in the cattle, supplemental feed costs, and miscellaneous expenses (COSTHD). Land and/or pasture costs (COSTHA) can be variable if charged on a per-head basis under a lease agreement, or fixed as in Eq. [7b] if the land is owned by the producer or charged as a flat charge per hectare as a part of a lease agreement.

The shape of the economic response to stocking rate can be seen in Fig. 9-2, which depicts both the individually computed values of NR for each of the treatments in the Sims et al. (1976) study and the theoretical NR surface predicted with Eq. [7]. Net returns in Fig. 9-2 were computed using the following values: PSELL = $1.35 kg^{-1}, PBUY = $1.49 kg^{-1}, WTBUY = 215 kg ha^{-1}, COSTHD = $55 head^{-1} (for a 150-d grazing period) and COSTHA = $18 ha^{-1}. PSELL and PBUY are estimated average beef prices based on the model of Schroeder et al. (1988) for Kansas markets for the period 1979 to 1988. Other expenses were estimated from Kansas stocker budgets prepared for the 1986 production year (McReynolds & Barnaby, 1986).

Figure 9-2 demonstrates that the NR response function of Eq. [7] is curvilinear (quadratic) when ADG is linear as in Eq. [4a]. Hence, an increase in SR from very low levels results in increased net returns until profits per hectare ($15.90) are maximized at a stocking rate of 0.97 steers ha^{-1}. Increases in SR beyond 0.97 head ha^{-1} will result in greater WTGNHA (Fig. 9-2), but profit will be less because the dollar value of those weight gains is not sufficient to compensate for the cost of purchasing and maintaining additional steers during the grazing season.

Figure 9-2 also shows that, depending on which variable is used to measure grazing system performance, there are three stocking rates that result in a *maximum* response: (i) very low SR (e.g., 0.20 head ha^{-1}) results in maximum ADG, WTGNHD, and sales weight per head; (ii) relatively high SR (e.g., 2.14 head ha^{-1}) results in maximum WTGNHA; and (iii) moderate SR (e.g., 0.97 head ha^{-1}) results in maximum NR or profits ($ ha^{-1}). From the producer's perspective, only one of these three maximizing SR levels is *optimal*, i.e., the one that maximizes NR.

Additionally, Fig. 9-2 reveals that even though 0.97 head ha^{-1} is the optimal SR for the producer who seeks to maximize predicted NR in the Sims et al. (1976) illustration, it will not remain optimal if any of the price/cost relationships change. Stated differently, use of the identical biophysical response surfaces for ADG, WTGNHD, or WTGNHA in conjunction with alternate values for any of the economic variables (PBUY, PSELL, or COSTHD) will result in an alternative predicted NR surface with specific slope, curvature, and maximum value. With either an increase in steer sale price PSELL, or a decrease in variable production expenses PBUY and COSTHD, the economically optimum SR will increase. By contrast, decreases in PSELL or increases in PBUY or COSTHD will result in an economic optimum at a lower SR. These impacts are discussed in detail in Hildreth and Riewe (1963).

STOCKING RATE AND RISK-RETURN TRADEOFFS

One management issue that the above analysis ignores is the weather risk associated with selecting a stocking rate. In a grazing system, the supply of herbage available to each animal (kg ha^{-1}) is a function of both the weather and the stocking rate. However, at the time the stockers are placed onto pasture at the outset of the grazing season, the producer lacks knowledge of whether rainfall over the grazing period will, in fact, provide sufficient

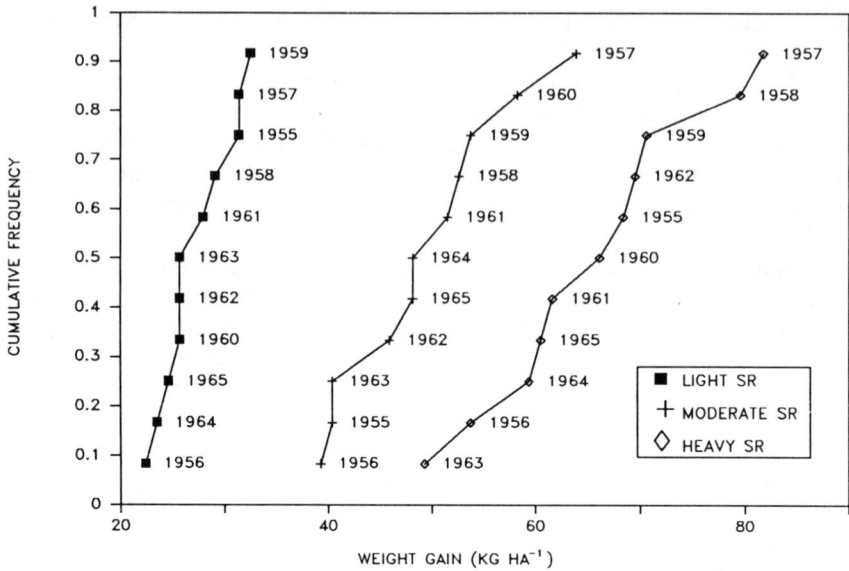

Fig. 9–3. Cumulative frequency (CDF) of WTGNHA at three steer stocking rates of range grasses.

pasture growth to support the optimal number of steers, which has been selected based on expected herbage availability. The continuum of possible mismatches between stocking rate and pasture availability *in any 1 yr* is bounded by the following two extremes: During a poor year, limited pasture availability could result in minimal weight gain per hectare (WTGNHA) due to excessive competition between animals. By contrast, during a good year, high levels of forage availability might result in large weight gains per head (ADG, WTGNHD) due to highly selective grazing but low returns per hectare (NR) due to excess ungrazed forage.

With year-to-year weather variability, the determination of the optimal SR is, at best, difficult. If perfect foreknowledge of weather and pasture availability exists, the producer would ideally match the stocking rate with the state of nature to ensure that there is sufficient herbage mass to support adequate weight gains and maximum net returns for that specific weather scenario. The producer might select, e.g., a high stocking rate in good years and a lower stocking rate in droughty years.

In the absence of this knowledge, however, the producer must select a stocking rate that is optimal in the sense that it gains a satisfactory level of net returns (NR) while concurrently minimizing exposure to weather risk. What constitutes a satisfactory level of NR as well as the level of risk a producer is willing and able to bear is both an individual and subjective matter that must be decided by the individual decision maker. To make this choice, however, the producer requires decision-support information that quantifies the risk-return tradeoffs associated with each stocking rate level. In essence, this information helps the producer to assess the following: (i) determine whether certain stocking rates are more susceptible to weather risk

ECONOMIC CONSIDERATIONS

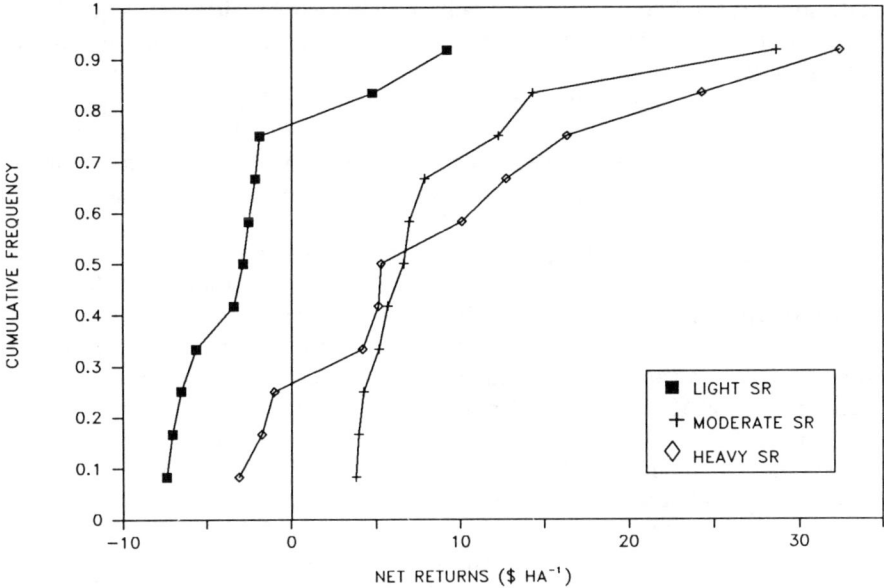

Fig. 9-4. Cumulative frequency (CDF) of NR at three steer stocking rates of range grasses.

(variability) than others; (ii) quantify how the weather variability affects animal and economic performance at each SR level; and (iii) assess the probability that any one state of nature—and its ensuing effect on weight gain and returns—will occur.

Various researchers have suggested that higher stocking rates result in greater risk for the producer (Harlan, 1958; Willms et al., 1986). Studies of steer grazing (White & Eidman, 1971; Curll, 1978) have shown that although high stocking rates result in superior economic performance in average or good years, that nevertheless, the adverse effects of a poor season were least for lower stocking rates.

The differential impact of weather risk at alternative stocking levels can be succinctly depicted with probabilistic estimates of animal response and economic performance. These probabilistic estimates are presented in the form of empirical cumulative distribution functions (Hogg & Tanis, 1977) for animal weight gain per hectare (WTGNHA) and economic returns (NR) for the Sims et al. (1976) grazing study in Fig. 9-3 and 9-4, respectively.

Probabilistic Estimates of Animal Response

One of the strengths of the experimental design of the Sims et al. (1976) study is that animal response to light, moderate, and heavy SR treatments was measured over an 11-yr period, 1955-1965. The light, moderate, and heavy SR treatments (0.25, 0.48, and 0.75 head ha^{-1}, respectively) resulted in a mean WTGNHA of 27, 49, and 65 kg ha^{-1} over this period with standard deviations (SD) of 3.25, 7.42, and 9.49 kg ha^{-1}, respectively. These data support the hypothesis that stocking rates that provide intensive forage utilization result in both increased animal gain per unit area and increased risk as measured by statistical variance.

Figure 9-3 presents this information from a probabilistic perspective.[2] The nearly vertical cumulative distribution function (CDF) for the light SR reflects the small variance in year-to-year WTGNHA. There is approximately a 0.10 probability that weight gains will be less than 22 kg ha^{-1} and a 0.90 probability that weight gains will be less than 32 kg ha^{-1}. Hence, in more than 8 yr out of 10 with this treatment, WTGNHA will vary within a narrow 10 kg ha^{-1} range despite the influence of annual weather variability. By contrast, the CDF for the heavy SR is less vertical with greater horizontal spread, indicating that animal gain per unit area is more susceptible to weather-induced variability associated with forage availability. However, because the CDF of WTGNHA for the high SR treatment lies entirely to the right of the CDF values for either the moderate or low SR treatments, it signifies that (i) at any specified probability level, WTGNHA will be higher under a high SR treatment; and (ii) at any specified WTGNHA level, there will be a greater probability of at least attaining weight gain per unit area at that level with a high SR. For example, at the 0.50 probability level (i.e., 5 yr out of 10), WTGNHA will be *at best* 50 kg ha^{-1} for the moderate SR vs. 68 kg ha^{-1} for the heavy SR. Conversely, the probability of attaining *at least* 50 kg ha^{-1} weight gain is 0.50 under the moderate SR compared to 0.90 with a heavy SR.

If weight gain response per unit area (WTGNHA) were the sole indicator of grazing system performance, then from the perspective of risk, the heavy SR treatment of the Sims et al. (1976) study would be preferred to other lower stocking rates. Although its year-to-year WTGNHA performance entails more variability as measured by its SD, the fact that its CDF lies entirely to the right of the other CDF values indicates that what Sims et al. (1976) classified as a high stocking rate resulted in greater absolute WTGNHA regardless of which state of nature occurred. From the perspective of animal gain per unit area (WTGNHA), the high SR treatment dominated the others.

Probabilistic Estimates of Economic Response

From the managerial viewpoint, with its emphasis on economic response as the sole, appropriate measure of grazing system performance, the optimal stocking rate in terms of risk-return implications must be identified using the CDF of NR for each of the three treatments as demonstrated in Fig. 9-4. Analogous to Fig. 9-3, as rangeland is stocked more heavily, the CDF values of net returns become increasingly less vertical and exhibit greater horizontal spread, which indicates greater variability of profits from year to year. However, unlike Fig. 9-3, the CDF of NR for the heavy SR treatment does not lie entirely to the right of the CDF for the moderate SR, and hence, does not unconditionally dominate it in terms of risk-returns.

---

[2]Cumulative probability levels in Fig. 9-3 were assigned using the fractile method described by Schlaifer (1959).

ECONOMIC CONSIDERATIONS

Although the CDF values of WTGNHA established that the Sims et al. (1976) heavy SR resulted in higher weight gains per unit area at all probability levels, these higher weight gains would not compensate for the higher costs associated with the heavy stocking rate during poor years when limited pasture availability would cause excess grazing pressure and minimal weight gains per head and per unit area. Figure 9-4 shows that in about 5 yr out of 10 (i.e., the 0.50 probability level), the heavily stocked range resulted in lower net returns than the moderate SR. Thus, the lower half of the CDF for the heavy SR lies to the left of the moderate stocking rate CDF for NR. Further inspection of these CDF values shows that NR with a heavy SR was negative 1 yr out of 4 (0.25 probability level), whereas even in the worst years, profits were realized on the moderately stocked range.

Mean net returns per hectare computed over the 11-yr period of the Sims et al. (1976) study were $-2.32, $9.03, and $9.49 for the light, moderate, and heavy SR treatments. Although the heavy SR rendered greater net returns *on average* than the moderate SR, the fact that there was a greater probability of earning low or negative returns during some years might influence a risk-averse producer to select the moderate SR as the optimal management alternative. In addition to forfeiting claim to higher average returns with a heavily stocked pasture, this producer simultaneously forgoes the increased probability of earning higher returns with a heavy SR during good years when competition for available pasture is not excessive. Stated differently, in an attempt to avoid the negative implications of the lower tail of the heavy stocking rate CDF, this risk-averse producer also sacrifices the positive implications of the upper end of this same CDF, which lies to the right of the CDF of the moderate SR treatment.

In cases where there are risk-return tradeoffs (i.e., higher returns bring higher risk) as demonstrated for the moderate and heavy SR treatments, the researcher cannot prescribe which stocking rate is optimal or best for the producer. However, research can provide the necessary probabilistic information contained in CDF values, which is prerequisite to identifying what those risk-return tradeoffs are. With this information, each grazing system manager can then individually identify the optimal stocking rate and its associated grazing pressure based on his willingness and ability to bear risk. For the above example, it is likely that the same information embedded in the CDF values of Fig. 9-4 would reveal other producers with either greater equity in their operation, sufficient cash savings, or lower family living expenses, who might be willing to venture with the high SR because of their increased ability to sustain low or negative returns over a short period.

IMPLICATIONS FOR DATA NEEDS AND DESIGN OF GRAZING STUDIES

The appropriateness of the experimental design and data collection in a field research study is ultimately determined by whether that design and data serve the objectives of the study. If the objective of the field researcher is, in part, to have the grazing study provide managerial guidelines from the economic perspective, then it is critical at outset to identify whether the proposed research design and data collection will permit the testing of economic hypotheses in a meaningful and relevant fashion. Under the assumption that producer economic objectives include (i) the maximization of net returns ($ ha^{-1}) and/or (ii) the appraisal of risk-return tradeoffs, several suggestions are offered that will enhance the potential for economic analysis of optimal stocking rates based on field grazing studies.

Weight Gain Response Surface

In analyzing the Sims et al. (1976) grazing study, the development of both the NR response surface (Eq. [7]) and the CDF of net returns (Fig. 9-4) hinged on the ability to predict animal weight gain response as a function of stocking rate. Hence, grazing study research design must incorporate data collection that permits the statistical estimation of weight gain response so that the implicit economic response surface of NR can be mathematically derived from it. Discussion of the rationale behind this judgement can be expressed in terms of three issues:

1. A weight gain response surface that is a function of SR characterizes an experimental design that corresponds to the input/output, cause/effect production conditions encountered by the producer in the managerial setting. With land as the fixed resource, the major *input* whose level the producer controls is the number of steers placed onto the pasture (SR). This fundamentally intuitive relationship of setting the SR at some fixed level to induce desirable weight gains is mirrored in the act of experimentally specifying treatments and factor levels to elicit a measurable response.

2. Weight gain as a function of SR implicitly acknowledges a continuous production response surface whose shape (e.g., slope, curvature) is disclosed only by setting the treatment (SR) at multiple levels. The NR function, which is used to determine the economically optimum stocking rate, is derived from the underlying weight gain response surface, and hence, is also a continuous function of SR. Unless the shape of this ensuing NR function is known, it is neither possible to determine where net returns ($ ha^{-1}) are maximized, nor is it possible to determine how the optimal stocking rate changes in response to fluctuations in prices and costs.

One shortcoming of many published grazing studies is that an insufficient number (e.g., two) of stocking rates are tested to see if one results in a significant increase in beef production over the other. To be sure, net returns can be calculated and compared at each treatment level analyzed. However, in the absence of a response surface of weight gain, it is impossible to determine if an optimum exists either at stocking rates in between the experimental levels, or at levels beyond those examined in the study. Hence, the design of grazing trials must include stocking rate treatment levels that encompass the potential range of grazing levels that could be economically optimal. Ideally, researchers would extend treatments at least to the level where WTGNHA reaches its maximum level to ensure that the field data bracket the area of economic relevance.

3. The necessity of multiple treatment levels over the broad range dictated by the response surface approach implicitly increases the cost of research by necessitating a larger number of observations. However, if there are limited resources for conducting the research, then for purposes of economic analysis, it would be preferable to reduce the number of replications at each treatment level (SR) in favor of having a larger number of treatment levels represented (Bransby, 1989).

Multiple-Year Data Collection

Analysis of the impact of weather variability on risk-return tradeoffs necessitates repetition of the grazing experiment over a sufficient number of years to capture the range and frequency of weather scenarios representative in the study area. In essence, the grazing experiment it replicated over time to estimate the year effect on grazing system response via variation in pasture availability. Critical to the design of a multiple-year study is that the same multiple stocking rate treatment levels be set each year to enable measurement of annual weight gain variations due to yearly changes in the grazing pressure at each SR level.

Results of a simulation study conducted by Parsch et al. (1987) illustrate this concept. Summer pasturing of yearling steers on common bermudagrass [*Cynodon dactylon* L. Pers.] were simulated over a 14-yr period (1972–1985) for each of eight stocking rates (between 2 head ha^{-1} and 16 head ha^{-1}) under western Arkansas conditions. Figure 9-5 shows how both wet (1979) and dry (1980) seasons affected simulated WTGNHA at each SR in comparison to a normal year. With ample precipitation, selective grazing

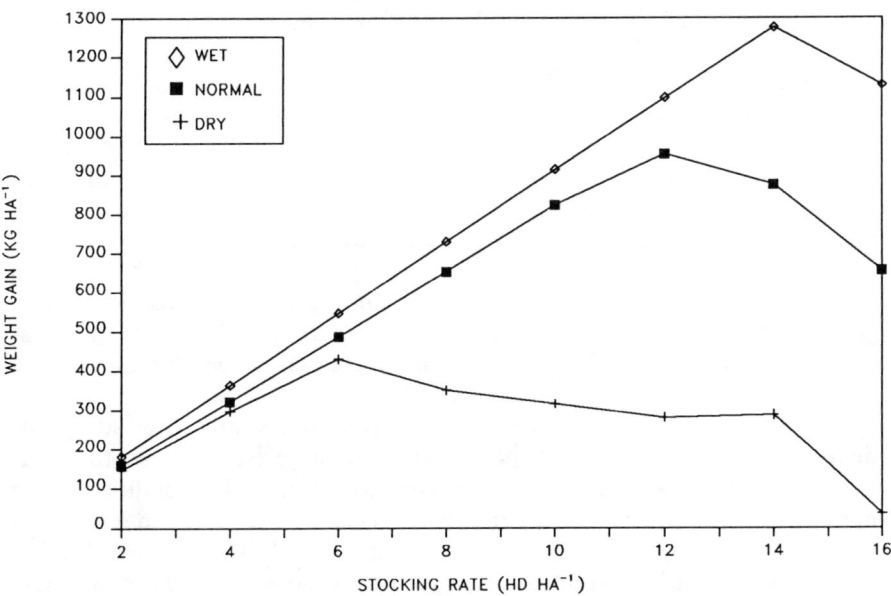

Fig. 9-5. Simulated WTGNHA for dry, normal, and wet seasons at eight stocking rates of yearling steers that grazed common bermudagrass.

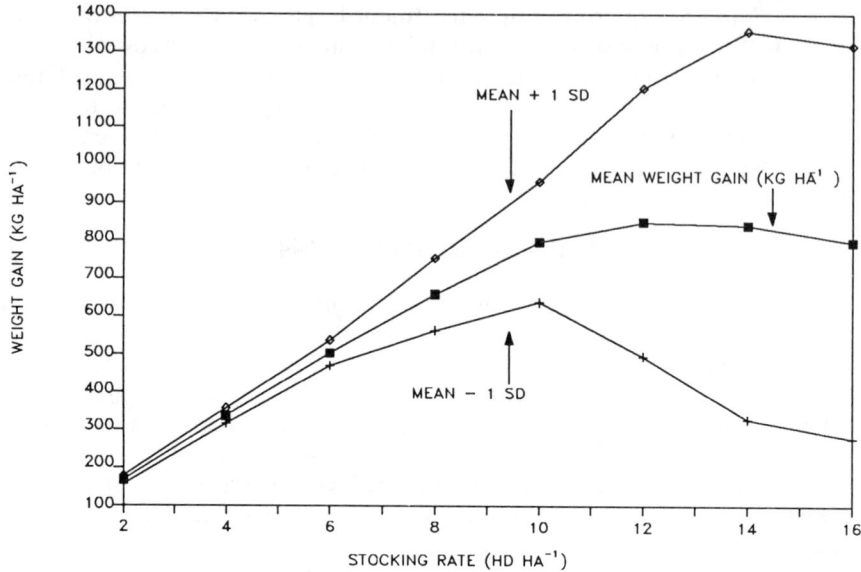

Fig. 9-6. Mean and standard deviation of simulated WTGNHA at eight stocking rates of yearling steers that grazed common bermudagrass.

was possible even at high levels of SR, resulting in maximum WTGNHA at 14 head. By contrast, during a dry year, grazing pressure and competition increased at all levels of SR resulting in reduced WTGNHA at high SR levels and maximum WTGNHA at a relatively low 6 head ha^{-1}.

Holding the SR levels constant from year to year for this simulation experiment enables computation of the standard deviation (SD) of WTGNHA at each SR. In Fig. 9-6, plots of the mean WTGNHA ±1 SD suggest increased variability and greater sensitivity to pasture availability at higher levels of SR. Mean simulated WTGNHA over the 14-yr period was maximized at 12 head ha^{-1} in the Parsch et al. (1987) study. Beyond 10 head ha^{-1}, risk—as measured by the SD—increased dramatically.

Figure 9-7 summarizes the corresponding economic values (NR) for the above study excluding pasture maintenance and land costs (COSTHA). Mean NR over the 14-yr simulation were maximized at 10 head ha^{-1} with progressively higher risk beyond 6 head ha^{-1} and increased probability of negative NR at stocking levels of 12 head ha^{-1} or more.

Multiple-year studies designed to assess risk implications place additional demands on the researcher in that the research objective includes the measurement of grazing system response under adverse or outlier conditions. One approach frequently used by field researchers is to throw out bad data when adverse weather results in yield loss or unusually low response. Because producers are faced with these same adverse conditions, the inclusion of these data in research publications is of value in that it demonstrates the potential risk inherent in that management alternative.

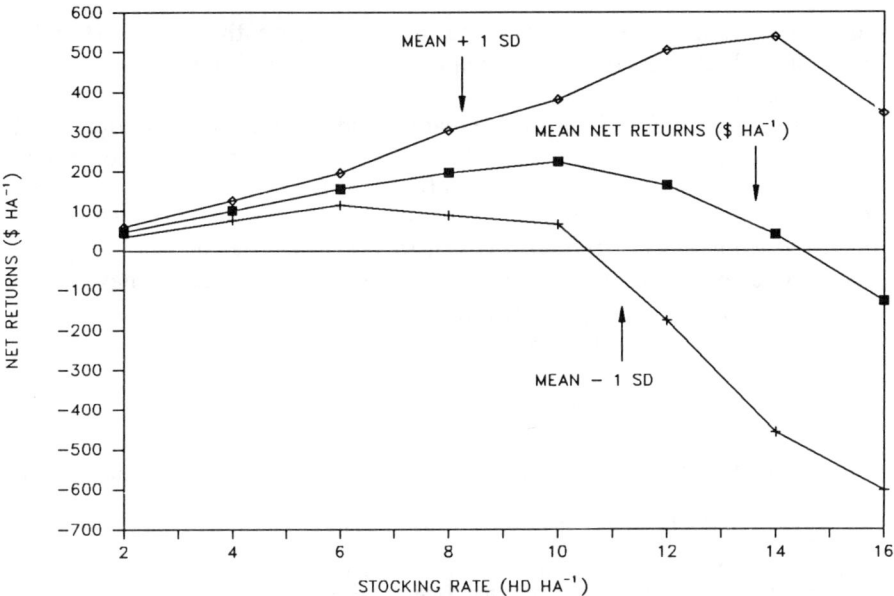

Fig. 9-7. Mean and standard deviation of simulated NR at eight stocking rates of yearling steers that grazed common bermudagrass.

In conducting multiple-year studies for risk analysis, the researcher must, at minimum, anticipate how the research design and data collection will be altered if adverse conditions become excessively severe. For example, low pasture availability during hot summer months may result in zero or negative weight gains for a high SR treatment. The researcher may elect to either continue the treatment, or alter it, or terminate it prematurely. If the researcher terminates the treatment, it should not result in thrown out data, because low weight gain response at these high grazing pressures provides a quantifiable measure of risk. Of course, premature termination by removing animals from the paddock may result in late-season forage growth if rain falls subsequent to their removal from pasture. Measurements of this unused forage will be necessary if the pasture has value as hay under this treatment. If, on the other hand, the researcher alters the treatment by providing, e.g., supplemental feed to the animals, the quantity and type of feed should be reported to enable economic cost accounting.

CONCLUSION

If the objective of the researcher is to conduct grazing studies that can be used as the basis for economic analysis, then the data that are collected must contribute to interpretation in a managerial, practical setting. In addressing the economic objective of maximizing profits or net returns, the most critical issue in grazing studies is to adequately define a response surface of weight gain as a continuous function of stocking rate based on multiple stocking rate treatment levels. In assessing the effect of weather risk on stocking rates, a study of similar design is required, except that it must be conducted over several years.

Although studies of this nature readily lend themselves to economic analysis and interpretation in that they reflect the relevant options and constraints faced by the producer, limited research funding, and resources may discourage researchers from undertaking multiple treatment grazing studies over a number of grazing seasons. In these circumstances, computer simulation experiments based on modeled relationships of livestock-forage systems may provide a suitable alternative to field studies.

REFERENCES

Adjei, M.B., P. Mislevy, and C.W. Ward. 1980. Response of tropical grasses to stocking rate. Agron. J. 72:863-868.

Bement, R.E. 1969. A stocking rate guide for beef production on blue-grama range. J. Range Manage. 22:83-86.

Bransby, D.I. 1989. Compromises in the design and conduct of grazing experiments. p. 53-67. In G.C. Marten (ed.) Grazing research: Design, methodology, and analysis. CSSA Spec. Publ. 16. ASA, CSSA, Madison, WI (Chapter 5 of this publication).

Bransby, D.I., B.E. Conrad, H.M. Dicks, and J.W. Drane. 1988. Justification for grazing intensity experiments: Analyzing and interpreting grazing data. J. Range Manage. 41:274-279.

Chapman, H.D., W.H. Marchant, P.R. Utley, R.E. Hellwig, and W.G. Monson. 1972. Performance of steers on pensacole bahiagrass, coastal bermudagrass, and coastcross-1 bermudagrass pastures and pellets. J. Anim. Sci. 34:373-378.

Conrad, B.E., E.C. Holt, and W.C. Ellis. 1981. Steer performance on coastal, callie, and other hybrid bermudagrasses. J. Anim. Sci. 42:1188-1192.

Curll, M.L. 1978. Simulation: An aid to decisions on superphosphate use for beef production. Ag. Syst. 3(3): 195-204.

Guerrero, J.N., B.E. Conrad, E.C. Holt, and H. Wu. 1984. Prediction of animal performance on bermudagrass pasture from available forage. Agron. J. 76:577-580.

Harlan, J.R. 1958. Generalized curves for gain per head and gain per acre in rates of grazing studies. J. Range Manage. 11:140-147.

Hart, R.H. 1972. Forage yield, stocking rate and beef gains on pasture. Herb. Abstr. 42:345-353.

Hart, R.H. 1978. Stocking rate theory and its application to grazing on rangelands. p. 547-550. In D.N. Hyser (ed.) Proc. 1st Int. Rangel. Congr., Denver, CO. 14-18 August. Society for Range Management, Denver, CO.

Hart, R.H., M.J. Samuel, P.S. Test, and M.A. Smith. 1988. Cattle, vegetation, and economic responses to grazing systems and grazing pressure. J. Range Manage. 41:282-286.

Hildreth, R.J., and M.E. Riewe. 1963. Grazing production curves. II. Determining the economic optimum stocking rate. Agron. J. 55:370-372.

Hogg, R.V., and E.A. Tanis. 1977. p. 83-105. In Probability and statistical inference. Macmillan Publishing Co., New York.

Hull, J.L., J.H. Meyer, S. Bonilla, and W. Weitkamp. 1965. Further studies on the influence of stocking rate on animal and forage production from irrigated pasture. J. Anim. Sci. 24:697-704.

Jones, R.J., and R.L. Sandland. 1974. The relation between animal gain and stocking rate. J. Agric. Sci. 83:335-342.

McReynolds, K.L., and G.A. Barnaby, Jr. 1986. Grazing yearling beef. Kansas State Univ. Coop. Ext. Serv. Farm Mgt. Guide MF-591.

Neville, W.E., and W.C. McCormick. 1976. Production of beef calves on coastal bermudagrass at two levels of grazing intensity: Costs and returns. J. Anim. Sci. 42:1404-1412.

Parsch, L.D., M.W. Watts, and O.J. Loewer. 1987. A risk-efficiency analysis of stocking rates for pasture-fed steers under weather uncertainty. Staff Pap. SP2187. Dep. of Agric. Econ., Univ. of Arkansas, Fayetteville, AR.

Peterson, R.G., H.L. Lucas, and G.O. Mott. 1965. Relationship between rate of stocking and per acre and per animal performance on pasture. Agron. J. 57:20-30.

Riewe, M.E. 1961. Use of the relationships of stocking rate to gain of cattle in an experimental design for grazing trials. Agron. J. 53:309-313.

Riewe, M.E. 1981. The economics of grazing. p. 517-526. *In* J.L. Wheeler and R.D. Mochrie (ed.) Forage evaluation: Concepts and techniques. Proc. Bilateral Workshop. Armidale, NSW, Australia. 27-31 Oct. 1980. Am. Forage and Grassl. Council, Lexington, KY.

Riewe, M.E., J.C. Smith, J.H. Jones, and E.C. Holt. 1963. Grazing production curves: I. Comparison of steer gains on Gulf ryegrass and tall fescue. Agron. J. 55:367-369.

Russell, K.D., R. Hironaka, and D.B. Wilson. 1981. An economic analysis of alternative management techniques for beef production on irrigated pastures in Alberta. Can. Farm Econ. 16:1-17.

Schlaifer, R. 1959. p. 103-104. *In* Probability and statistics for business decisions. McGraw-Hill, New York.

Schroeder, T., J. Mintert, F. Brazle, and O. Grunewald. 1988. Factors affecting feeder cattle price differentials. West. J. Agric. Econ. 13:71-81.

Sims, P.L., B.E. Dahl, and A.H. Denham. 1976. Vegetation and livestock response at three grazing intensities on sandhill rangeland in eastern Colorado. Colorado State Univ., Agric. Exp. Stn. Tech. Bull. 130.

Torell, L.A., and R.H. Hart. 1988. Economic consideration for efficient stocking rates on rangeland. p. 71-76. *In* R.S. White and R.E. Short (ed.) Proc. Symp. Achieving Efficient Use of Rangeland Resources, Miles City, MT. 14-16 Sept. 1987. Montana State Univ. Agric. Exp. Stn. and USDA-ARS, MilesCity, MT.

White, F.C., and V.R. Eidman. 1971. The bayesian decision model with more than one predictor—an application to the stocking rate problem. South. J. Agric. Econ. 3:95-102.

Willms, W.D., S. Smoliak, and G.B. Schaalje. 1986. Cattle weight gains in relation to stocking rate on rough fescue grassland. J. Range Manage. 39:182-186.

Workman, J.P., and J.M. Fowler. 1986. Optimum stocking rate biology vs. economics. p. 101-102. *In* P.J. Ross et al. (ed.) Proc. 2nd Int. Rangel. Congr.: Rangeland Under Siege, Adelaide, South Australia. 13-18 May 1984. Australian Academy of Science, Canberra, A.C.T., Australia.

10 Issues on Modeling Grazing Systems

Otto J. Loewer
University of Arkansas
Fayetteville, Arkansas

ABSTRACT

We are all modelers of one type or another. Researchers who conduct grazing trials are utilizing/developing physical models that are biologically complete but mathematically incomplete. Researchers who use/develop dynamic simulations are utilizing/developing mathematical models that are mathematically complete but biologically incomplete. All models should be examined in terms of their objectives, assumptions, completeness, sensitivity, credibility, and ability to predict while recognizing that not all models are created equal. Ideally, grazing trial researchers and researchers who utilize dynamic simulation should work together in establishing a mutually beneficial grazing experiment. Researchers coordinate their efforts with statisticians to enhance the quality of the experiment. A similar procedure is recommended for the use of dynamic simulations.

Plant and animal scientists who conduct field-grazing experiments do not often utilize computer simulation modeling. Agricultural scientists who develop computer simulation models do not often directly conduct grazing experiments. Given this situation, the *objectives* of this chapter are as follows:

1. Comment on modeling of biological systems
2. Discuss modeling as a means of extending results from grazing trials
3. Identify essential plant and animal data that needs to be acquired from grazing trials

MODELING PHILOSOPHY

We are all modelers. Each of us observes the world that surrounds us. We formulate how we believe it functions in terms of physical, social, and economic relationships. This type of model is called a *mental model* and is composed of images and words that we use in communicating our under-

---

Copyright © 1989 Crop Science Society of America and American Society of Agronomy, 677 S. Segoe Rd., Madison, WI 53711, USA. *Grazing Research: Design, Methodology, and Analysis*, CSSA Special Publication no. 16.

standing to others. When we make an observation that cannot be explained by our existing mental model, we reformulate it taking into account the added information. The process of model restructuring is called *learning*.

Mental models may be used to form physical models, the term *physical* being used in its broadest sense to include physics, biology, and chemistry. A physical model is an object or system made to scale, a simple example being a model airplane. Grazing trials, however, may also be classified as physical models in that they attempt to represent scaled-down grazing systems.

Mathematical models represent another type of model. Mathematical models result when mental models are expressed in mathematical form, thus quantifying our description of the system in question. Mathematical models may be placed into the following categories: statistical, mechanistic, and simulation.

Statistical models include regression models. Regression models utilize experimental observations to mathematically relate one or more input parameters with an associated output. An example of a statistical regression model is the net energy system that predicts average daily gain of beef cattle (*Bos* sp.) given dry matter intake and energy content (Lofgreen & Garret, 1968).

Mechanistic models are based on a mathematical form that is believed to best represent the relationship between one or more input parameters and an associated output. Input terms in the mathematical expression may often be conceptual rather than physical. For example, a power function may be used to describe the weight-age growth relationship (Brown et al., 1976). Thus, for the equation $Y = aX^b$, the exponent of the mathematical function (b) has conceptual rather than physical significance.

Simulation models most often utilize computers for computational purposes and, hence, are referred to as computer models. *Computer models* are collections of mathematical-logical expressions that relate inputs and outputs. These expressions may include both statistical and mechanistic relationships. Optimization models, such as linear programs, are sometimes referred to as *static simulations*. Models that predict relational changes that occur over time are called *dynamic simulations*. Dynamic simulation models offer the greatest potential for describing the complex relationships associated with grazing trials.

OBSERVATIONS ABOUT MODELS

Modeling is used extensively in plant and animal research including grazing trials. Most often, however, the models are relatively simple and are statistical or mechanistic rather than dynamic computer simulations. For example, the net energy system (Lofgren & Garrett, 1968) and ADF-NDF concepts (Van Soest, 1982) are widely accepted models consisting of a few equations.

All models are not created equal. This is especially true for dynamic computer simulations. The sufficiency of a given model, be it mental, physical, or mathematical, is dependent on how well it addresses the following concerns: objectives, assumptions, completeness, sensitivity to inputs, credibility of the modeler, and ability to predict.

Objectives

Each model has its own stated or implied objectives. That is, what is it that the model is intended to do, and who is the intended user? Once these objectives are known, it is possible to evaluate the suitability of the model's underlying assumptions.

Assumptions

Every model has specified or implied assumptions. Some factors are considered sufficiently important to the objectives that they are treated separately, whereas all others are grouped together in what statisticians call the *error term*. Assumptions are used to establish the completeness of the model.

Completeness

Completeness refers to the degree to which a process is described within a model. For example, the net energy system is a statistical regression model that predicts weight gain given a quantity and energy level of feed consumption. Inherent to this model are stated and implied assumptions as to genotype, weather, minerals, health, etc. The net energy model does not directly address the biochemistry associated with weight gain. Hence, it would be considered incomplete if adding insight into biochemical reactions were among the reasons for using the model. Similarly, the model would be incomplete if one wished to evaluate the effects of temperature on weight gain unless the maintenance aspects of energy utilization were addressed separately. However, the model would be considered sufficiently complete if the objectives were to predict weight gains given feed energy inputs, and none of the assumptions associated with the initial development of the model were violated.

Sensitivity to Inputs

Models are evaluated, consciously or subconsciously, by the degree to which they are sensitive to inputs. In either grazing trials or dynamic simulations, for example, it might be of interest to determine the sensitivity of total beef production to changes in stocking rate, other factors remaining constant. The level of credibility of both model forms depends on the extent to which our expectations are satisfied. For example, there would be little confidence in a model that predicted ever-increasing total beef production (weight/unit area) for ever-increasing stocking rates (animals/unit area). We would place more confidence in a model that predicted an initial increase in total beef production followed by a decrease in that the model reflects what has been observed in earlier studies.

Credibility of the Modeler

A major component of model building is the model builder. Respect for a model often depends on how honestly and openly the model builder describes his model. Model builders, be they of grazing trials or computer simulations, enhance their models through complete documentation of their efforts.

Ability to Predict

In the final analysis, models are judged mainly by their ability to predict. A model may fail to predict because of the following:

1. Initial status of the system is incorrectly defined (for example, the initial soil fertility input levels are incorrectly given)
2. Functional relationships are not specified correctly (for example, grazing animals require more energy from forage than was initially believed)
3. Timing and/or level of changes to the system are incorrectly given (for example, fertilizer is applied at a different time or in different amounts than was specified)
4. Combinations of the above

Thus, a model may be conceptually correct and yet lack the ability to predict with acceptable degrees of accuracy. Regardless, grazing trial models are ultimately evaluated by farmers and ranchers who utilize the results from these models in making management decisions. Computer simulation models are judged more critically because their results can be completely quantified

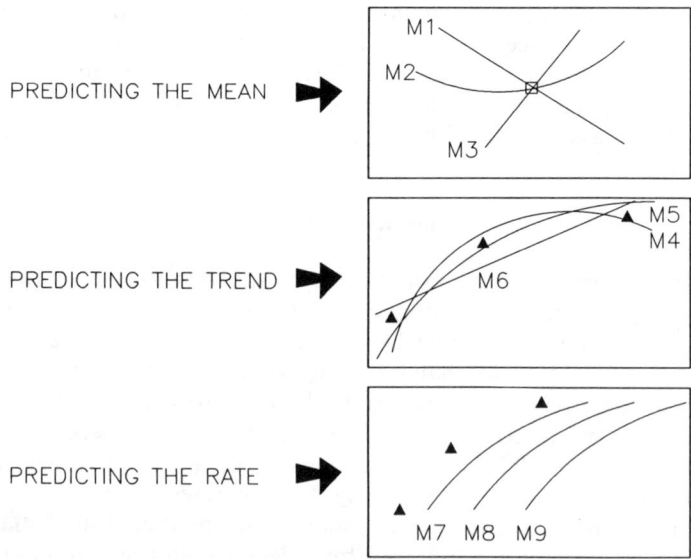

Fig. 10-1. Comparative predictability standards for models.

and reproduced for comparative purposes. There are several standards by which a model might be compared in terms of measuring predictability (Fig. 10-1). For example, three different models might predict the mean but arrive at that point very differently. Likewise, models may differ in their ability to predict trends or rates of change. Generally, computer simulations are more accurate in predicting trends and rates of change rather than mean values. Grazing trials and other field experiments, however, tend to report a limited number of mean values for purposes of comparison, the inference often being that the discrete points may be linearly connected.

REASONS FOR MATHEMATICAL MODELING

Researchers develop mathematical models for several reasons, including:

1. Mathematical models are perhaps the only way to quantitatively describe complex biological × physical × chemical × economic interactions over time
2. Mathematical models can recreate history relatively quickly by using appropriate input records (such as weather) to answer the "What if . . . ?" kinds of questions
3. Mathematical models can generate information continuously through simulated time, effectively defining intermediate values between laboratory and/or field observations

GRAZING TRIALS AND COMPUTER SIMULATION MODELS

Although grazing trials and computer simulations are different types of models, they can be used effectively to complement and enhance each other. A greater understanding of biological and mathematical completeness by researchers developing each type of model can lead to a greater understanding of the grazing system.

Mathematics and Biology

A grazing trial is a form of physical model. It is biologically complete in that all the biological systems are functioning during the trial. However, it is mathematically incomplete in that not all (in fact, very few) of the biological systems are mathematically quantified over the course of the study (Fig. 10-2). Conversely, a dynamic computer simulation model is mathematically complete in that one or more equations are used to describe the grazing system. However, the simulation model is biologically incomplete in that not all (in fact, very few) of the individual biological processes are described mathematically. It is as unlikely that all biological components of the physical grazing system will be quantified over time as it is that they will be described mathematically in a dynamic computer simulation. However, it is important to recognize that grazing trials and computer simulation are both forms of modeling that differ in their completeness.

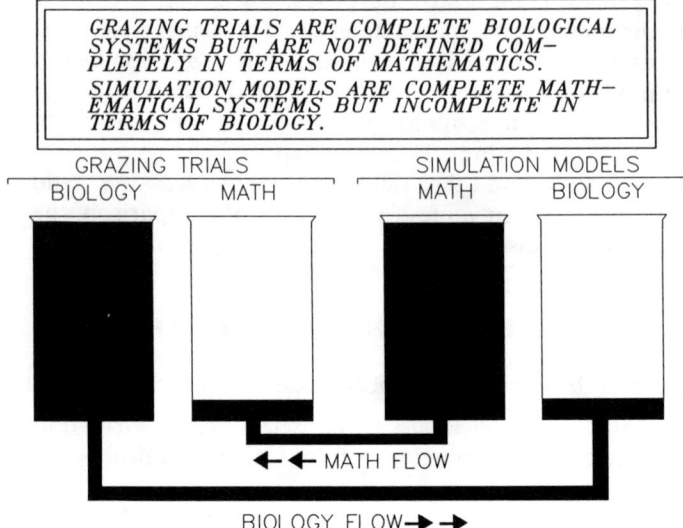

Fig. 10-2. Comparative completeness of grazing trials and mathematical models.

Field Comparisons

Grazing trial results are usually viewed with more confidence than similar results obtained from dynamic simulations. The question "Has the model been validated?" is often asked, and rightly so. Upon further examination, however, the concept of validation is considerably more complex than might first be believed. For example, validation infers that there is some known standard of performance by which comparisons may be made. Grazing trial data usually becomes the defacto standard. Yet, a grazing trial is another form of model. It is biologically complete but mathematically incomplete. What usually happens is that the results from a dynamic simulation (mathematically complete but biologically incomplete) are compared to relatively few mean values from a grazing trial. Numerous mathematical assumptions have to be made about the grazing trial and then incorporated into the dynamic simulation. So much is unknown that the comparisons are of limited value; thus, the term *field comparison* is perhaps more appropriate than validation.

For purposes of illustration, Fig. 10-3 is a field comparison of results from the GRAZE model (Loewer et al., 1987) and a grazing trial. The grazing trial had been conducted independently without anticipating that it would be used for comparative purposes. The comparison was initiated in response to the request of the grazing trial researchers who had an interest in GRAZE but wanted to see if it were valid for their location. GRAZE seems to have performed very well. However, let us examine the comparison more closely in terms of objectives, assumptions, completeness, and predictability.

Although GRAZE can simulate any genotype, the best body composition values by far are for shorthorn steers (Moulton et al., 1922). In addi-

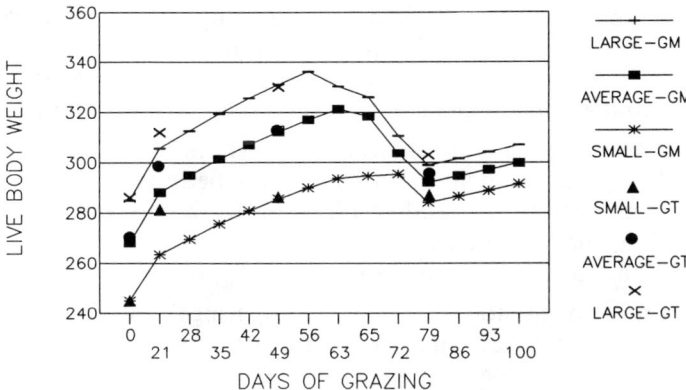

Fig. 10-3. Field comparison.

tion, only one type of animal can be simulated at a time in GRAZE. The grazing trial used a cross between Angus and an exotic genotype. Also, steers and heifers were grazed together. Thus, an equivalent-steer concept was used in GRAZE to account for stocking rate, and the body composition assumptions based on shorthorn steers were considered adequate.

GRAZE reports empty body rather than full body weights. The grazing trial measured beginning and ending shrink weights; that is, the animals were weighted after a number of hours of being isolated from feed sources. However, intermediate weights were obtained without allowing time for shrink. Thus, there was little consistency in the ways the weights were obtained, and the fraction of live body weight that is empty body weight can only be estimated. In addition, only four measurements of body weight were made over the 80-d grazing period, and forage growth, availability, intake, and quality were not measured.

The grazing trial did provide performance data for individual animals. Initial weight differences between the largest and smallest animal were approximately 15%. The average animal more closely resembled the larger animal. Animals were fed hay of undetermined quality and quantity after Day 80 of the trial.

Weather records were obtained from the nearest official weather station rather than on-site. No record of relative animal health was provided. GRAZE version 1.0 treats each pasture as a monoculture with bermudagrass [*Cynodon dactylon* (L.) Pers.] being used for the comparison. In the grazing trial, the pasture was a mixture of bermudagrass and clover (*Trifolium* sp.).

Figure 10-3 compares 12 field measurements of animal weight to animal weight values resulting from thousands of mathematical calculations used in GRAZE based on numerous assumptions and varying degrees of biological completeness. A formal statistical analysis might indicate the extent to which results satisfy at least one definition of a validated model. However, because of the many assumptions that had to be made in converting grazing trial inputs to simulation model inputs, it is difficult to say much more than that GRAZE predicted body weight reasonably well. In addition, the implied inference of a valid model is that its many internal components must also be valid. For example, if GRAZE accurately predicts weight gain, then the many different submodels that influence weight gain must also be correct. Of course, this might not be true because of the following:

1. There may be compensating error within the model
2. Not all components of a model may be tested
3. Sensitivity of the model to its range of inputs and input combinations may not be known
4. Combinations of the above

Thus, in my opinion, exercises in comparing results from grazing trials and dynamic simulations are desirable and necessary. However, attaching a valid or not valid label to a model should be tied closely to the test(s) and assumptions used for validation.

GRAZING TRIAL DATA AND SIMULATION MODELS

Common Ground

Researchers who utilize dynamic simulation models are often asked by researchers who conduct grazing trials "What data is essential for me to collect so that my study can be used to validate your model?" Sometimes the question really is "How can my grazing trial data be used more effectively, and can it be used in your simulation model?" These are legitimate questions. The degree to which they are answerable depends, in part, on the following:

1. Researchers who conduct grazing trials and researchers who build, test and utilize simulation models share a common interest in trying to unlock the secrets of grazing systems. Thus, each group can enhance its own efforts by working as a coordinated team in examining points of common interest.
2. Again, all models are not created equal. Thus, for maximum utilization of grazing trial data in simulation models, the grazing trial researcher should identify the simulation model that he wishes to use prior to actually conducting the study. Researchers do this type of planning with statistical models and should do the same with simulation models.
3. Most simulation models are dynamic in that they predict changes in the status of the system over time. Conversely, grazing trial data as reported in the literature tends to reflect long-term averages making it less usable by modelers in making field comparisons. Remember, on the average, all things are average including results from both grazing trials and computer simulation models.

Essential Data from Grazing Trials

The term *essential data* begs the question "Essential for what?" Most often, essential data refers to physical measurements that form the minimum set of information needed to provide for some acceptably realistic comparison between a field study and a computer simulation model. The inference is that the grazing trial is to be conducted independently from any computer simulation model. Because every simulation model is different, there is no unique set of essential data. However, the following types of information are common to most models, although the degree of complexity varies greatly.

Initial Status of the Grazing System. Information needed concerning the initial status of pastures includes: proportions and types of species, dry matter accumulations and quality, soil types, and soil test data. Information needed concerning the initial status of grazing animals includes: stocking rate, age, genotype, sex and reproductive status (if applicable), animal weight (measured systematically and individually), and health.

Dynamic Status of the Grazing System. There needs to be frequent, consistent, and defined measurements of the items given in the previous paragraph. Local daily weather information is needed, the minimum being rainfall and minimum and maximum temperature. Detailed management schedules and descriptions of cultural practices should be included such as fertilizer applications (rates, dates, and types) and animal rotations (times and numbers moved).

Voids in Essential Data

Voids in essential data are dependent on the specific model being used. For example, the net energy system does not use body composition. Thus, body composition information is not a void for this particular model. In general, however, dynamic computer simulation models have need for additional information concerning the following:

Animals. More information is needed to better describe physiological age and body composition relationships as functions of genotype, sex, and reproductive status.

Plants. More information is needed to enhance our understanding of the relationships among plant physiological age and physical and chemical attributes, especially as related to a quantifiable measure of plant quality.

Animal Health. There is a great void in the whole area of animal health in terms of defining growth and reproduction. Specifically, we need to examine the roles of parasites and diseases.

Plant-Animal Interface. The greatest challenge to simulation modeling is in predicting intake. For example, the types of approaches used by simulation modelers to predict dry matter intake include percent of body weight, feed energy intake maximum, internal chemostatic control, physical fill limitations, feed chemical composition, physical characteristics of feed, number and size of bites, social behavior, and rest requirements.

Each of these factors probably impacts intake under some conditions. The challenge is to appropriately incorporate them into models and measure them in grazing trials. Thus, the greatest need is to determine the fundamental relationships involved in defining animal intake and utilization.

SUMMARY

In summary, we are all modelers of one type or another. Researchers who conduct grazing trials are utilizing/developing physical models that are biologically complete but mathematically incomplete. Researchers who use/develop dynamic simulations are utilizing/developing mathematical models that are mathematically complete but biologically incomplete. All models should be examined in terms of their objectives, assumptions, completeness, sensitivity, credibility, and ability to predict while recognizing that not all models are created equal. Ideally, grazing trial researchers and researchers who utilize dynamic simulation should work together in establishing a mutually beneficial grazing experiment. Researchers coordinate their efforts with statisticians to enhance the quality of the experiment. A similar procedure is recommended for the use of dynamic simulations.

REFERENCES

Brown, J.E., H.A. Fitzhugh, Jr., and T.C. Cartwright. 1976. A comparison of nonlinear models for describing weight-age relationships in cattle. J. Anim. Sci. 42:810–818.

Loewer, O.J., K.L. Taul, L.W. Turner, N. Gay, and R. Muntifering. 1987. GRAZE: A model of selective grazing by beef animals. Agric. Systems 25:197–309.

Lofgreen, G.P., and W.N. Garrett. 1968. A system for expressing net energy requirements. J. Anim. Sci. 27:793–806.

Moulton, C.R., P.F. Trowbridge, and L.D. Haigh. 1922. Studies in animal nutrition: III. Changes in chemical composition on different planes of nutrition. Univ. of Missouri Agric. Exp. Stn. Bull. 55.

Van Soest, P.J. 1982. Nutritional ecology of the ruminant. O&B Books, Corvallis, OR.